# LIFE, DEATH AND NITRIC OXIDE

# RSC Paperbacks

RSC Paperbacks are a series of inexpensive texts suitable for teachers and students and give a clear, readable introduction to selected topics in chemistry. They should also appeal to the general chemist. For further information on all available titles contact:

Sales and Customer Care Department, Royal Society of Chemistry,
Thomas Graham House, Science Park, Milton Road, Cambridge CB4 0WF, UK
Telephone: +44 (0)1223 432360; Fax: +44 (0)1223 423429; E-mail: sales@rsc.org

Recent Titles Available

**The Chemistry of Fireworks**
*By Michael S. Russell*
**Water (Second Edition): A Matrix of Life**
*By Felix Franks*
**The Science of Chocolate**
*By Stephen T. Beckett*
**The Science of Sugar Confectionery**
*By W.P. Edwards*
**Colour Chemistry**
*By R.M. Christie*
**Beer: Quality, Safety and Nutritional Aspects**
*By P.S. Hughes and E.D. Baxter*
**Understanding Batteries**
*By Ronald M. Dell and David A.J. Rand*
**Principles of Thermal Analysis and Calorimetry**
*Edited by P.J. Haines*
**Food: The Chemistry of Its Components (Fourth Edition)**
*By Tom P. Coultate*
**The Misuse of Drugs Act: A Guide for Forensic Scientists**
*By L.A. King*

Future titles may be obtained immediately on publication by placing a standing order for RSC Paperbacks. Information on this is available from the address above.

RSC Paperbacks

# LIFE, DEATH AND NITRIC OXIDE

ANTHONY BUTLER

*University of St Andrews, Fife, Scotland*

ROSSLYN NICHOLSON

*Stavanger, Norway*

advancing the chemical sciences

ISBN 0-85404-686-0

A catalogue record for this book is available from the British Library

Published by The Royal Society of Chemistry,
Thomas Graham House, Science Park, Milton Road,
Cambridge CB4 0WF, UK
Registered Charity Number 207890

For further information see our web site at www.rsc.org

Typeset by H Charlesworth & Co Ltd, Huddersfield, UK.
Printed by TJ International, Padstow, Cornwall, UK

# Preface

There are not many small molecules about which a whole book could be written. Rather unexpectedly, at least to those who studied chemistry and biology before 1987, the simple diatomic molecule nitric oxide (NO)* is one such species. That it progressed from being a nasty pollutant in the air we breath to being designated 'molecule of the year', because of its biological role, by the journal *Science* in 1992 has been said and written so many times that it would be tedious to dwell on this again. That said, it is still a remarkable story.

The dramatic change in our perception of NO came about when it was discovered that it is part of mammalian physiology, firstly as a messenger molecule in controlling the dilation of blood vessels and subsequently in a whole host of tissues and organs. The story is still unfolding. The literary theme is not inappropriate as the scientist-turned-author Carl Djerassi has written a novel entitled *Nitric Oxide*. However, there was an 'NO story' before its physiological role was discovered: it has an interesting chemistry and plays a crucial part in the nitrogen cycle in soil and in smog formation. In an effort to rid the world's cities of smog the chemistry of NO was fully explored by the oil companies as they struggled to find ways of reducing NO levels in car exhausts to keep ahead of the demanding legislation that was being enacted. Much of that work was, of course, commercially sensitive and received little publicity. The news that NO plays a part in controlling blood flow, and hence blood pressure, was publicized through scientific journals and brought fame, and a Nobel prize, to some of those involved. What has been truly amazing is the subsequent proliferation of roles for NO in animal physiology. As this subject has been under intensive scrutiny by scientists for over 150 years it is astonishing that no-one before had seriously suggested that animals make NO as part of normal physiology. There is, to our knowledge, not even a hint of this

---

* Although not strictly correct usage, it is customary to refer to nitric oxide as NO, using the chemical formula as an abbreviation. Biologists often put a dot after the formula to indicate that it is a radical but we have not felt this to be necessary.

in the pre-1987 scientific literature. Perhaps the reasons are rather prosaic: NO is a very small molecule, it has a very short lifetime in tissue (how short depends on whom you ask but it rarely exceeds 30 s) and it is difficult to detect chemically at the concentrations at which it occurs (less than 1 µM).

In this book we have written about many areas of chemistry and biology and, in order to keep the book to a reasonable length, some topics have been greatly simplified. This could lead to misunderstandings and we trust that experts in these areas will not be too critical. At the same time we do apologize for any errors and omissions, and hope that the message of the book ('Look at what one simple, diatomic molecule can do') comes through in spite of our shortcomings. Shakespeare described Cleopatra as having 'infinite variety' and the same might be said of NO. That infinite variety is a result of its very special chemistry and we hope that we have written something that makes sense of the NO story in terms of its chemistry.

When NO was first publicized as a cellular messenger molecule, some biologists saw that it is a radical (a species with an unpaired electron) and assumed that it must be highly reactive, like other biologically significant radicals. Chemists, of course, knew that NO, although a radical, is not particularly reactive. If you understand chemical bonding the reason for the lack of reactivity is easy to understand but try explaining it to a cardiologist in a hurry. From school chemistry we may know that NO reacts readily with oxygen to give brown fumes of $NO_2$. Open a gas jar of NO to the atmosphere, an act now probably forbidden by EU regulations, and the reaction appears to be instantaneous. However, at the concentrations of NO and oxygen found in tissue the reaction is much slower, a nice example of the law of mass action. Immunologists are not always sensitive to the finer points of reaction kinetics and it is not easy to explain why the formation of $NO_2$ is generally unimportant to the biological role of NO. For a time there were problems discussing NO with some medical doctors because of confusion with nitrous oxide ($N_2O$), with which they were familiar because of its place in anaesthesia ('laughing gas').* Now that inhaled NO is used therapeutically let us hope that this confusion has been overcome.

We have not written this book because the NO story is now coming to an end but, as so much progress has been made, it is not a bad time to take stock. Anyone who has written a scientific review knows that

---

* Sir Humphry Davy discovered laughing gas but its effect on respiration was first investigated by a physician, Peter Mark Roget, later to achieve greater fame as a compiler of a thesaurus of the English language.

one of the most demanding and time-consuming aspects of such writing is keeping track of all the references. We have refrained from giving references in the text, partly to make the text easier to read and partly to shorten the book. Our selection of names for mention has been arbitrary. Scientific research can appear very arid and occasionally mentioning a person may remind readers that without people there is no science.

Much of the NO story has still to be fully elucidated. This means that the current status of our understanding in some areas is one of confusion and doubt, particularly concerning the biological role of NO. A recent review of NO in the skin* was subtitled *More questions than answers*. We have tried to avoid emphasizing the confusion because it weakens the impact of the message. Instead we thought that in this preface we would warn the reader that some of the conclusions are stated with a confidence they do not strictly deserve. There are two major sources of confusion: firstly, results obtained from experiments using cultured cells may not apply to whole animals and, secondly, the effect of NO in tissue is often reversed if the concentration is raised. If the conclusions are seen as interim ones then little harm can be done. At the end of each chapter we have given references for further reading for those who wish to know more.

Readers of this book require some knowledge of chemistry and biology but we have tried to explain in the text as many technical terms as possible. A glossary of technical terms has been provided for further help. Some chapters will be easier for the general reader to understand than others. These are probably those at the end of the book rather than at the beginning. Some chapters, such as Chapter 8 and 9, have been included for reference and serious study rather than casual reading.

We have failed to cover all aspects of the chemistry and biology of NO. For example, it plays a crucial role in the development of insulin-dependent diabetes and in influenza and asthma. Also we have only touched on the fascinating co-ordination chemistry of NO. Had we tried to mention everything the book might have been better but it would never have been finished. An issue of the American Chemical Society publication *Chemical Reviews* (April 2002) has a comprehensive account of many aspects of the chemistry of NO by experts and repays study by the serious student of the subject.

Before we conclude we would like to thank some people. We thank the RSC for inviting us to write this book; we hope it will not regret its

---

* R. Weller, Nitric oxide, skin growth and differentiation. *Clin. Exptl. Dermatol.*, 1999, **24**, 388.

decision. To all those scientists who had the insight to see that NO could do all the things we have tried to describe, and then laboured long and hard in their laboratories to prove their ideas, we tender our heart-felt thanks. We thank a number of experts who read what we have written and made helpful suggestions for improvement: Richard Weller (Edinburgh Royal Infirmary), Ian Megson (University of Edinburgh), David Adams (Heriot Watt University), Keith Sillar (University of St Andrews), William Martin (University of Glasgow), Faisel Khan (Ninewells Hospital, Dundee), Roberta Fruttero (University of Torino), David Cole-Hamilton (University of St Andrews), Lyn Williams (University of Durham), Neil Cape (Institute of Terrestrial Ecology), John Moffett (Needham Research Institute) and Kevin Nolan (Royal College of Surgeons in Ireland). We owe a special debt of gratitude to Eric Flitney (University of St Andrews), for introducing us to the biology of NO in exchange for some information on the photochemistry of nitroprusside.

Finally we thank our spouses, Janet and Hugh, who encouraged us to start the work and rejoice with us at its completion.

Anthony Butler
*St Andrews, Scotland*

Rosslyn Nicholson
*Stavanger, Norway*

# Contents

# Glossary

**Activation energy**. The amount of energy required by the reacting molecules to form an activated complex when they collide and thus begin a chemical reaction. If two colliding molecules do not possess enough energy, they bounce apart again and the collision does not result in a reaction.

**Adrenergic.** A nerve in which the neurotransmitter is norepinephrine (noradrenaline).

**Anaemia (anemia).** A diseased state characterized by an inadequate supply of haemoglobin.

**Angina.** Narrowing of coronary arteries due to atherosclerosis, restricting the supply of oxygen to the heart.

**Antibonding orbital.** A molecular orbital higher in energy than any of the atomic orbitals from which it is derived. When an antibonding orbital is populated with electrons, the molecule becomes less stable.

**Aorta.** A major blood vessel from the heart.

**Apoptosis.** Cell death in which cells are programmed to die.

**Bond order.** A theoretical index of the degree of bonding between two atoms, relative to a two-electron covalent bond, *e.g.* C–H.

**Calmodulin.** A calcium ion containing protein essential for the action of NOS.

**Carcinogen.** A substance that can provoke the growth of cancerous cells.

**Cholinergic.** A nerve in which the neurotransmitter is acetylcholine.

**Cyclic guanosine-3, 5-monophosphate (cGMP).** Substance formed as muscles relax, Scheme 1.1.

**Cytokine.** A substance that stimulates the immune system.

**Dipole moment.** An expression of the degree of polarity in a molecule.

**Disproportionation.** Any chemical reaction whereby two molecules of one chemical react together to form two different chemical species, *i.e.* $A + A \rightarrow B + C$.

**Electrophile.** Any species that is attracted to a negative centre.

**Endothelial cells.** Cells forming the endothelium (see below).

**Endothelium.** A single layer of thin plate-like cells lining the inside of a blood vessel.

**Endothelium-derived relaxing factor (EDRF).** Messenger molecule that stimulates vascular muscle to relax.

**Entropy.** A measure of randomness or disorder within a system, usually denoted by the letter $S$.

**Enzyme.** A very large molecule (generally a protein) that catalyses vitally important chemical reactions occurring in living systems.

**Glial cells.** Cells in the brain that support the nerve cells.

**Glyceryl trinitrate.** Drug used for the treatment of angina (Formula 1.1).

**Ground state.** The lowest stable energy state of a system, such as a molecule or atom.

**Guanosine-5-triphosphate (GTP).** Substance consumed as muscles relax, Scheme 1.1.

**Guanylate cyclase (sGC).** Enzyme responsible for, *inter alia*, muscle relaxation.

**Half-life.** The time taken for the concentration of one reactant in a reaction to fall to half its original value.

**Hybrid orbital.** An electron orbital formed *via* combination of two or more orbitals of different shapes, *e.g.* s and p orbitals on an atom.

**Hydrophobic.** Molecules or parts of molecules that are insoluble in water. Such molecules and groups tend to be non-polar.

**Immune system.** The body's defence against foreign substances.

***In vitro.*** A reaction occurring outside a living system (*lit.* in glassware).

***In vivo.*** A reaction occurring in a living system.

**Ionization potential.** A measure of the tendency of an atom or molecule to form a positive species *via* loss of an electron.

**Isoform.** One member of a family of enzymes.

**Ligand.** In inorganic chemistry a ligand is an atom or group of atoms bound to a central atom. In biochemistry the term has been used more widely; wherever a group of atoms may be regarded as a central unit, any atom or group of atoms bound to it may be referred to as a ligand.

**Lumen.** The hollow part of a blood vessel.

**Macrophage.** A mobile scavenger cell, part of the immune system. Protects the body against invading microbes.

**Mass spectrometry.** A very sensitive technique for measuring the relative molecular mass (molecular weight) of a chemical species.

**Micro-organism.** Any organism that can be viewed only with the aid of a microscope. Micro-organisms include bacteria and viruses amongst others.

**Mitochrondrion.** A rod-like structure within a cell. It controls certain activities of the cell, particularly respiration.

**Necrosis.** Cell death induced by toxic chemicals.

**Neurotransmitter.** Chemical that carries a signal from one nerve cell to another across a synapse.

**Nitrergic.** A nerve in which one of the neurotransmitters is NO.

**Nitric oxide synthase (NOS).** A family of very large enzymes that bring about the production of NO from the naturally occurring amino acid arginine. There are at least three members of the family, named eNOS, nNOS and iNOS.

**Nucleophile.** Any species that is attracted to a positive centre.

**Peroxynitrite.** A highly reactive ion formed by the reaction of NO and superoxide. It rapidly isomerizes to nitrate.

**Phagocytosis.** A process of engulfing and destroying bacteria by cells of the immune system.

**Platelet.** Subcellular structure found in blood, important in the formation of clots.

**Platelet aggregation.** Clumping of platelets to form a plug.

**Postsynaptic.** The nerve cell coming after the synapse.

**Presynaptic.** The nerve cell coming before the synapse.

**Prostagladins.** A group of physiologically active substances with a range of functions, including the prevention of platelet aggregation.

**Psoriasis.** Chronic inflammatory skin disease affecting almost 2% of the population.

**Respiration.** The metabolic process in animals and plants in which organic substances are broken down into simpler molecules with the release of energy. This usually (but not exclusively) requires oxygen.

**Shear stress.** The mechanical stress caused as blood flows over the inner surface of a blood vessel.

**Smog.** An unpleasant, acrid form of fog caused by industrial pollution.

**Smooth muscle.** A type of muscle found throughout the bodies of many animals, called smooth because of its appearance under the microscope. It is the major component of the walls of blood vessels.

**Stenosis.** An unnatural narrowing of any passage in the body, particularly blood vessels.

**Stent.** A surgical device to keep open a passage that has undergone stenosis.

**Synapse.** The anatomical relationship of one nerve cell to another.

**Tumour.** Strictly any swelling but generally restricted to swellings due to uncontrolled cell growth.

**Valence shell.** The outermost electron shell of an atom containing electrons in the ground state.

**Valency.** The bond-forming capacity of an atom, expressed either as the number of single bonds it can form with other atoms or the number of electrons an atom gives up or accepts when forming a bond.

**Vasculature.** The system of vessels conducting blood through the body.

*Chapter 1*

# What on Earth is Nitric Oxide Doing Here?

The year 1952 was a bad one for Londoners. The city was frequently shrouded in smog – something between a smoke and a fog – causing traffic chaos and considerable physical distress, particularly amongst the more elderly members of the population. Smog forms when the moisture in a fog, for which London had long been famous, condenses around tiny particles from industrial emissions. Then sulphur dioxide, coming from the burning of coal rich in sulphur in open fires, dissolves in the moisture to give the acrid taste so characteristic of smog. The authorities took rapid action to reduce the amount of smoke and sulphur dioxide released into the air in urban areas. London's smog episodes were quickly eradicated.

London was by no means the only city with smog problems. For some years similar measures had been taken in Los Angeles to reduce smoke and particulate emissions, but the chronic smog problem had failed to ease. It transpired, much to the surprise of the local residents, that the Los Angeles smog had a different cause. Car exhaust fumes, in which the air above Los Angeles abounded, contained not only carbon dioxide and water but also traces of two oxides of nitrogen (nitric oxide, NO, and nitrogen dioxide, $NO_2$) as well as volatile organic compounds. To complicate matters further, the strong Californian sunlight provided suitable energy for a complex series of reactions involving NO which gave a 'photochemical smog'. This was just as distressing as London's industrial smog. Exposure caused eye and bronchial irritation in humans, it blanched the leaves of trees and it accelerated the corrosion of rubber. This is the context in which, before 1987, NO was usually mentioned.

NO is a simple gas and each molecule contains one atom of nitrogen and one atom of oxygen. It just gets mentioned in a school chemistry syllabus. Its most well-known property is that when it is released into the atmosphere it reacts with oxygen to form a brown gas, nitrogen dioxide $NO_2$. This means that if NO forms in the atmosphere the result is a mixture of NO and $NO_2$. Formation of NO does occur naturally, but only in the extreme conditions of a lightning strike. For convenience the mixture of NO and $NO_2$ is known as NOx, pronounced to rhyme with 'socks'. Areas with high NOx concentrations (and hence photochemical smog) sometimes appear on weather charts to warn people to keep away if they can.

Without NOx photochemical smog does not form and so cars in Los Angeles, and elsewhere, are now fitted with catalytic converters to destroy the NOx before it enters the atmosphere. This process is described in more detail later (Chapter 7). With increasing understanding of air pollution NO and $NO_2$ were seen as major villains in the saga and great efforts were made to banish them. However, in 1987 scientists were astonished to learn of another side of NO's complex character: human life depends upon it.

Vessels that carry blood around the body can enlarge, or dilate, and this fixes the amount of blood that is delivered to specific tissues or organs. To see how this happens we have to look at the structure of a blood vessel. The hole in the middle, down which the blood flows, is called the lumen. On the inside of the lumen is a single layer of cells known as the endothelium (endothelial cells) and the wall of the vessel consists largely of smooth muscle (Figure 1.1). It is called smooth

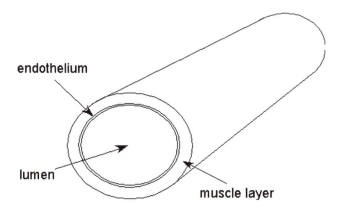

**Figure 1.1**   *Structure of an artery showing the endothelium*

because of its appearance under the microscope. When the muscles of the artery wall relax the lumen enlarges and more blood flows through the vessel, provided the heart is pumping properly. If the muscles contract the lumen decreases in size. Then, either less blood flows along the vessel or the heart has to work harder to maintain the flow of the same amount of blood. This is why 'hardening' of the arteries, which prevents muscle relaxation, puts strain on the heart.

It had been known for many years that certain substances, such as acetylcholine, bring about vascular muscle relaxation and some of these substances are released into the bloodstream when relaxation is required. During the late 1970s this matter was under investigation by one of the world's leading muscle physiologists, Robert Furchgott (Figure 1.2), working at the Downstate Medical Center in New York City. He wanted to know how and why substances like acetylcholine affected muscles in this way and he was working with small segments of artery taken from a rabbit. He and his student (John Zawadzki) encountered the most trying of situations in scientific research; their results were not reproducible. In one set of experiments acetylcholine was an active muscle relaxant but in another set of apparently identical experiments it had hardly any relaxing effect at all. The most natural thing to do when this happens is to give up and go fishing, but Furchgott and Zawadzki were made of sterner stuff and they eventually concluded that the result they obtained depended on whether or not the endothelium was intact. It is extremely easy to damage the endothelium when the experiment is being set up.

If the endothelium was undamaged acetylcholine worked fine; remove the endothelium by rubbing, either intentional or accidental, and acetylcholine had no effect at all. Furchgott concluded, quite rightly it is now clear, that acetylcholine was not acting directly on the muscle cells but on the endothelial cells which, in turn, produced another chemical species that diffused into the surrounding muscle and began the process of relaxation. Such a substance is called a messenger molecule as it tells the muscles what to do. Furchgott named this particular messenger molecule the 'endothelium-derived relaxing factor'. It was seen as important enough to warrant a set of initials (but not an acronym): EDRF. The chemical identity of the EDRF was a matter of intense study by many scientists during the 1980s. It was assumed to be a complex organic molecule, like most messenger molecules in the body, but it stubbornly resisted identification. You can see the magnitude of the challenge. The EDRF is produced, along with hundreds of other chemicals, by endothelial cells in quantities around

**Figure 1.2** *Robert Furchgott (left) receiving a Nobel prize from the King of Sweden*

a thousand millionth of a gram, well below the limits of detection by normal chemical means.

The problem was solved by an inspired guess and, although the first person to make the suggestion in print was Furchgott himself, the same idea occurred to others. As is described in more detail in Chapter 5, for many years drugs have been available for bringing about relaxation of the muscle in blood vessels. One, glyceryl trinitrate (**1.1**), is used either as a lozenge or in a puffer by people suffering from angina. Another one, sodium nitroprusside (**1.2**), is used in the management of cardiac crises to relax vascular muscle and so lessen the workload on the heart.

$$\left[\begin{array}{l} -ONO_2 \\[1em] -ONO_2 \\[1em] -ONO_2 \end{array}\right.$$

1.1

$$\left[\begin{array}{c} CN \\ NC_{\prime\prime\prime\prime\prime\prime}\!\!-\!\!Fe\!\!-\!\!{}^{\prime\prime\prime\prime\prime}CN \\ NC \qquad\qquad NO \\ CN \end{array}\right]^{2+}$$

1.2

The action of both drugs was discovered by accident and both now have valued places in medical practice. Why they act as muscle relaxants was, in the 1980s, not known but it was generally assumed that, along with many clinical drugs, their mode of action was completely different from that of naturally occurring vascular muscle relaxants, but perhaps this was not correct. Possibly both glyceryl trinitrate and sodium nitroprusside are transformed within the body into something that is also the naturally occurring messenger molecule for effecting vascular muscle relaxation, the mysterious EDRF. Both drugs contain a group of atoms that includes nitrogen and oxygen and so, the speculation went, the EDRF *could* be NO. With hindsight the idea is fairly obvious but at the time it was considered absurd. No-one had ever detected NO, or anything like it, in a living system. However, once the suggestion had been made several groups of researchers set about testing it. Two groups were successful in showing that the EDRF is indeed NO. One was led by Louis Ignarro (Figure 1.3), working at the University of California Medical School in Los Angeles and the other was led by Salvador Moncada (Figure 1.4), then at the Wellcome Research Laboratories in Beckenham, England.

They both used a bioassay to establish the identity of the EDRF. What this amounts to is a series of experiments in which a solution of NO and a solution of the EDRF were shown to have identical characteristics with respect to the relaxation of vascular muscle. Thus the evidence, initially, was circumstantial but quite quickly Moncada and his group produced direct evidence for the production of NO from cultured endothelial cells. This is described in more detail later (Chapter 3). Accounts of the work were published simultaneously by the two groups in January 1987. One paper appeared in the journal *Nature* and the other in *Proceedings of the National Academy of Sciences of the USA*. The publications were greeted with excitement and some incredulity. Excitement because of their importance in understanding muscle physiology, and incredulity because the findings were was so unexpected. The EDRF was not the complex organic molecule we had expected but a small, diatomic, *inorganic* molecule. The unexpected nature of the discovery led many to express doubt. NO is produced in lightning strikes and in internal combustion engines, not in mammals at

**Figure 1.3** *Louis Ignarro, the American scientist who identified the EDRF as nitric oxide*

body temperature. Anyway, it is a harmful substance, responsible for photochemical smog, not a benign substance keeping our cardiovascular system in a healthy state. Also, many said that NO is too reactive to function in a living system. It reacts with oxygen to produce $NO_2$, which is highly toxic. None of these objections was sustained on further investigation and it is now firmly established that NO is the EDRF and responsible for vascular muscle relaxation. When sufferers take glyceryl trinitrate for relief of the symptoms of angina the drug is converted into NO, which causes the cardiac arteries to enlarge and more blood to flow. The significance of the discovery was recognized by the award of the 1998 Nobel Prize for Physiology or Medicine, the highest award in the world of science, to three scientists involved in advancing our understanding of muscle relaxation, Ferid Murad, Robert Furchgott and Louis Ignarro. The last two were directly involved in the discovery of NO. Astonishingly the award did not include Salvador Moncada, a matter that has been commented upon publicly on a number of occasions. The papers of the Nobel Committee are not made public for

**Figure 1.4** *Salvador Moncada, the South American born British scientist who identified the EDRF as nitric oxide*

50 years, long after the deaths of most of those who are so puzzled and disappointed by this omission.

Had that been the end of the NO story it would have been a remarkable discovery, adding another piece to the jigsaw puzzle that makes up the chemistry of life. However, it was only the beginning. This small, diatomic, inorganic molecule turns out to have so many functions in living processes that we wonder how we missed it for so long. At the same time, NO is still polluting the atmosphere and more and more cars are being fitted with catalytic converters to rid the air of this dangerous chemical. Unlike almost any other molecule NO has two diametrically opposed characteristics, giving us both life and death.

It is now firmly established that NO is a cellular messenger molecule bringing about vascular smooth muscle relaxation, but exactly how does it do so? Muscle relaxation is a positive process rather than merely the absence of contraction. It is accompanied by the conversion, within the cell, of guanosine-5-triphosphate (GTP) into cyclic guanosine-3,5-monophosphate (cGMP), shown in Scheme 1.1. This

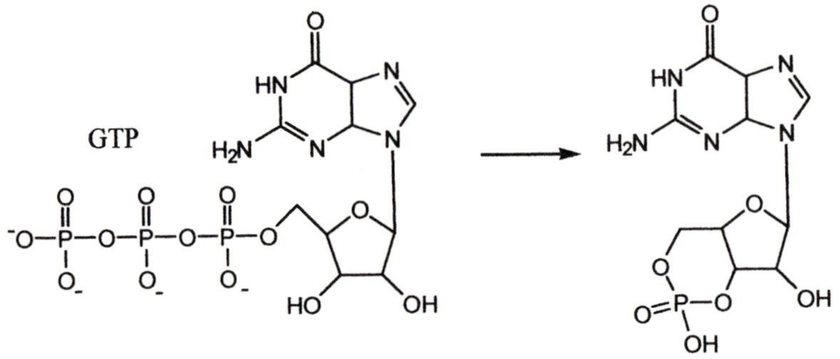

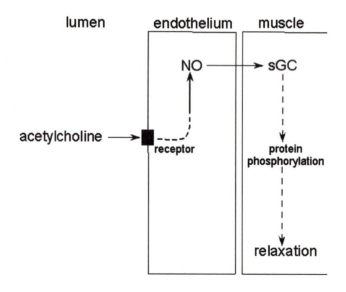

**Scheme 1.1** *Reaction resulting from the activation of guanylate cyclase*

conversion initiates a whole cascade of reactions, leading to protein phosphorylation (attachment of phosphate to a hydroxyl group) and muscle relaxation, processes well understood but too complex to be detailed here. It suffices to know that conversion of GTP into cGMP is catalysed by the enzyme-soluble guanylate cyclase (sGC, also known as guanylyl cyclase), an enzyme activated by NO.

The process observed by Furchgott is shown in Figure 1.5. Acetylcholine binds to a receptor on the surface of an endothelial cell

lumen          endothelium     muscle

NO ⟶ sGC

acetylcholine ⟶ ▪
receptor

protein
phosphorylation

relaxation

(– – – indicates sequence of reactions)

**Figure 1.5** *NO as the messenger molecule in arterial muscle relaxation*

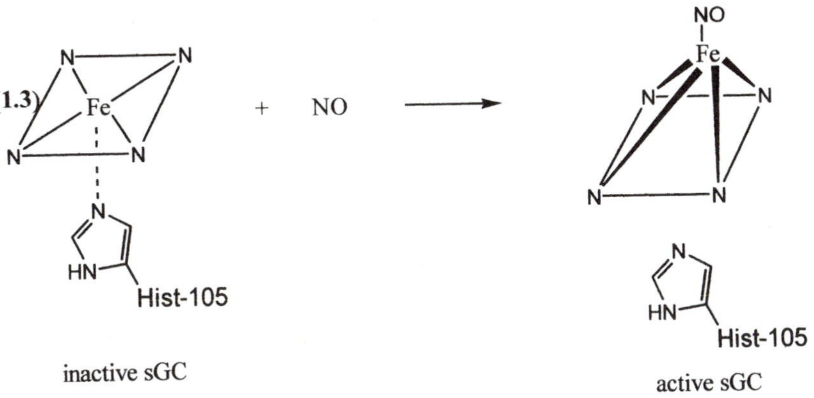

**Scheme 1.2**  *How NO activates guanylate cyclase by binding to iron in the haem component*

and, in a manner described in more detail later, stimulates the cell to produce NO, which then diffuses into the underlying muscle cell and activates sGC to bring about relaxation. Destroy or damage the endothelium and acetylcholine will have no effect.

sGC is found in the liquid part of almost all mammalian cells, with the highest concentration in the cells of the brain and lungs. It is described as soluble GC because it is in solution and there is another form, particulate GC, that is found in membranes. At the time of writing the three-dimensional structure of sGC has not been determined but the enzyme is known to contain a haem component. Haem is the non-protein part of haemoglobin and consists of an iron atom held in the centre of a porphyrin ring (**1.3** in Scheme 1.2). The protein part of sGC is a dimer of two similar but not identical subunits, named $\alpha$ and $\beta$. It is the haem component that makes sGC sensitive to NO. If the haem component is removed, as may happen during isolation and purification, the sensitivity of sGC to NO is lost. The haem component is bound to the enzyme *via* an imidazole ligand (part of histidine at position 105) in the $\beta$ subunit. Spectral evidence suggests that the iron is present as $Fe^{2+}$ bound to imidazole, as in nonoxygenated haemoglobin. The latter readily binds both oxygen and NO. In contrast, sGC, because of structural features in the $\alpha$ subunit, has a remarkably low affinity for oxygen yet readily binds NO. If it were not for this distinction, sGC in cells would be oxygenated and useless as a receptor for NO.

When NO binds to the haem iron of sGC, the bond to the histidine at position 105 (which is *trans* to the incoming NO) is broken (Scheme

1.2). This results in the formation of a pentaco-ordinated nitrosyl haem complex; the reason for this is explained in Chapter 9. At the same time the iron moves out of the plane of the porphyrin ring, which exposes the catalytic site of the enzyme and allows GTP to enter. Catalytic conversion of GTP into cGMP is dependent on the presence of $Mg^{2+}$ ions as this cation binds to a number of negatively charged groups in the enzyme and gives the required rigidity to the three-dimensional structure. The other metal known to be necessary for sGC activity is $Cu^+$. Animals fed on a copper-deficient diet exhibit marked loss of endothelium-dependent smooth muscle relaxation. The role of $Cu^+$ ions is unclear but, as described in Chapter 4, $Cu^+$ ions strongly catalyse the release of NO from *S*-nitrosothiols and so nitrosation of thiols on sGC *may* be a step in the activation of the enzyme. From this brief account of just one enzyme it is apparent how much inorganic chemistry there is in the chemistry of living processes.

sGC is a constitutive enzyme in that it is present nearly everywhere in animal cells all of the time. In conditions such as septic shock (see Chapter 11), where there is overproduction of NO, it is important to know how far elevated levels of sGC are also responsible for the massive arterial dilation that leads to life-threatening loss of blood pressure. Drugs to inactivate sCG, as well inhibitors of the enzyme responsible for the production of NO (see Chapter 3), may be relevant to the treatment of this condition. The same consideration applies to other faults in the cardiovascular system. Raynaud's phenomenon ('cold finger syndrome') is a troublesome condition that is caused by very restricted flow in peripheral blood vessels, particularly in the fingers. In extreme cases the sufferer cannot go out in cold weather without electrically heated gloves. It is, for some reason, much more common amongst young women than men and tends to lessen with age. The simplest explanation of the effect is that it is due to a shortfall in the production of NO and there is insufficient dilation of minor arteries to allow warming blood to reach the fingers. This condition could be cured by applying NO-releasing drugs to the skin (transdermal delivery of NO), and this is quite feasible. However, experimental evidence suggests that there is also a shortage of sGC in the fingers of sufferers, a situation far more difficult to relieve by means of drugs. Much more research on sGC is required if we are to appreciate its role not only in vascular smooth muscle relaxation, but also in the other living processes in which it plays a part. The elucidation of its activation by NO has done much to stimulate interest in sGC and surely it must be one of the next enzymes to yield its secrets.

# FURTHER READING

R.F. Furchgott and J.V. Zawadzki, The obligatory role of endothelial cells in the relaxation of arterial smooth muscle by acetylcholine. *Nature*, 1980, **288**, 373.

R.M.J. Palmer, A.G. Ferrige and S. Moncada, Nitric oxide release accounts for the biological activity of endothelium-derived relaxing factor. *Nature*, 1987, **327**, 524.

L.J. Ignarro, G.M. Buga, K.S. Wood, R.E. Byrns and G. Chaudhri, Endothelium-derived relaxing factor produced and released from artery and vein is nitric oxide. *Proc. Natl. Acad. Sci. USA*, 1987, **84**, 9265.

Y.-C. Lee, E. Martin and F. Murad, Human recombinant soluble guanylyl cyclase: expression, purification, and regulation. *Proc. Natl. Acad. Sci. USA*, 2000, **97**, 10763.

A.J. Hobbs, Soluble guanylate cyclase: the forgotten sibling. *Trends in Pharm. Sci.*, 1997, **18**, 484.

J.J.F. Belch and M. Ho, Pharmacotherapy of Raynaud's phenomenon. *Drugs*, 1996, **52**, 682.

J. Leppert, Å. Ringqvist, J. Ahlner, U. Myrdal, S. Sörensen and I. Ringqvist, Cold-exposure increases cyclic guanosine-monophosphate in healthy women but not in women with Raynauds-phenomenon. *J. Intern. Med.*, 1995, **237**, 493.

T. Gura, Eyes on the prize. *Nature*, 2001, **413**, 560. http://www.nobel.se/medicine/laureates/1998

*Chapter 2*

# Stopping Clots

Blood is pumped around the body at fairly high pressure. If it were not for a very sophisticated repair mechanism even a tiny leakage from a damaged blood vessel would result in the victim bleeding to death, so the prevention of bleeding is vitally important to our well-being. Loss of blood from a very small artery, vein or capillary is readily dealt with. The initial response is constriction of the sheath of smooth muscle and this slows down the flow of blood to the damaged area. It also results in opposing endothelial surfaces of the vessel being pressed together. A 'stickiness' develops, holds them together and may close off the vessel permanently. With larger vessels permanent closure is not acceptable and blood loss is prevented by other means involving blood platelets.

Blood platelets are cell fragments that circulate in blood. They are 2 μm in diameter, much smaller than either red or white cells. In healthy blood there are 250 million platelets in each millilitre of blood and they originate, as do other blood components, from stem cells found in bone marrow. Platelets are not particularly structured but contain numerous granules. They have a tendency to adhere to the endothelium but, under normal circumstances, this does not occur, for reasons to be explained shortly. If the endothelium of a blood vessel is damaged, and the underlying tissue exposed, platelets adhere to the exposed surface and the attachment causes them to release a chemical mixture stored in the granules. This chemical mixture makes further platelets adhere to those already attached and, eventually, a plug of platelets is formed. This process is known as platelet aggregation. A substance known as the von Willebrand factor, which originates in the endothelium, facilitates the adherence of platelets to the vessel wall. Any abnormality in the production of this factor has serious consequences for the process of aggregation. Platelet aggregation is further stimulated by the production of thromboxane $A_2$ from the platelet (Figure 2.1).

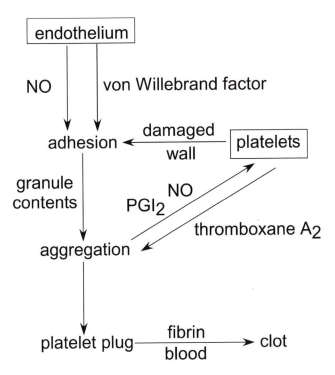

**Figure 2.1**   *Some of the processes involved in the formation and prevention of blood clots*

Bleeding is thus prevented by the platelet plug completely sealing the break in the vessel wall but if aggregation were to continue it would eventually completely block the vessel. This is prevented by the release from the vessel wall adjacent to the growing plug of a chemical agent that inhibits platelet aggregation called prostaglandin $I_2$ ($PGI_2$). When the platelet plug grows large enough to make contact with undamaged endothelium, $PGI_2$ brings the process of aggregation to a halt. The creation of the platelet plug occurs very rapidly and blood loss is stopped, but the repair is fragile. Stronger repair necessitates a second process, blood coagulation, which is much slower. Blood plasma contains a protein, fibrinogen, and this is transformed by an enzyme produced at the damaged site into a protein polymer known as fibrin, a mesh of interacting strands. Further enzyme action strengthens the three-dimensional structure and many blood cells become trapped within the fibrin matrix until it becomes a solid gel, termed a clot or thrombus. The entrapped cells are not an essential part of the coagulation process but entrapment commonly occurs. The role of the clot is to bolster the blocking action of the platelet plug. Eventually the damaged vessel re-grows, the clot and plug disintegrate and all returns to normal. If thrombus formation occurs when it is not required the situation is serious and anti-clotting drugs may be administered. One of these is, of course, aspirin.

Although, as mentioned earlier, platelets have a natural tendency to adhere to surfaces and to one another this, in a healthy person, does not occur. The reason is that endothelial cells produce substances that are inhibitors of platelet aggregation and the opposing forces of aggregation and anti-aggregation are finely balanced. One of these inhibitors, $PGI_2$, has been mentioned already as being responsible for limiting the size of the platelet plug. However, soon after the discovery of prostaglandin it became apparent that its generation and release accounts for only part of the anti-aggregating ability of the vascular endothelium. Even if prostaglandin generation is completely inhibited, endothelial cells can still inhibit platelet aggregation. Even before the EDRF was identified as NO it was suggested that the EDRF might act as an inhibitor of platelet aggregation. Once NO had been established as the EDRF it was easy to see whether or not NO inhibits platelet aggregation and this was found to be the case. Thus, not only does the arginine-to-NO pathway in endothelial cells provide the messenger molecule for smooth muscle relaxation, it also prevents the unwanted and inappropriate aggregation of platelets. It is now easy to see why damage to the endothelium results in adhesion and aggregation. It also explains why drugs such as glyceryl trinitrate are inhibitors of platelet aggregation; such substances are converted into NO *in vivo*. The endothelial cells produce most of the NO involved in inhibition of aggregation, but platelets themselves produce a little NO and there may be a source of NO in blood plasma (an *S*-nitrosothiol, see Chapter 4). The processes involved in the activation and inhibition of platelet aggregation and clot formation are summarized in Figure 2.1.

Platelets are uniformly suspended in blood plasma and for the NO produced by endothelial cells to reach them it must first dissolve and then diffuse through the plasma. Haemoglobin, contained in red blood cells, reacts rapidly with NO (it binds to the iron atom in haemoglobin in much the same way as to the iron atom in guanylate cyclase) and there is plenty of blood to soak up all the NO produced. How then does endothelial NO ever reach platelets? The answer lies in the behaviour of blood when flowing. The flow is streamed with all the solid particles and, in particular, the red blood cells concentrated in the centre while next to the vessel wall there is a zone free of red blood cells. However, as platelets are much smaller, they are not streamed in the same way and there are plenty in this region. It is these platelets which form a plug if the endothelium is damaged. In a healthy vessel $PGI_2$ and NO from endothelial cells prevent adhesion and aggregation but the NO would not reach platelets if it were not for the flow characteristics of blood. If blood flow becomes turbulent, rather than streamed, the red-cell-free zone is lost and unwanted platelet aggregation is more likely to occur. A partial blockage of a vessel, caused by the laying down of cholesterol-rich atherosclerotic plaque, leads to turbulence. This may be one reason why a high cholesterol diet is a cause of cardiovascular disease.

How does NO prevent platelet adhesion and aggregation? As with smooth muscle, platelets contain guanylate cyclase and activation of this enzyme by NO results in conversion of GTP into cGMP and initiates a cascade of reactions affecting the concentration of calcium ions in various parts of the platelet. Why this should influence the stickiness of the platelets remains to be discovered. One of the first groups of researchers to identify a role for NO in the inhibition of platelet adhesion and aggregation was Salvador Moncada and his colleagues at the Wellcome Research Laboratories. As described in Chapter 1, Moncada was also one of the first to identify NO as the EDRF, but that is not all. It was he, while working with Sir John Vane, who recognized $PGI_2$ as an endothelium-derived inhibitor of platelet aggregation. It is a remarkable record.

If a thrombus or clot forms in a blood vessel when it is not the response to bleeding, the situation is serious as blockage may result. One treatment is the administration of drugs that break down the clot. Now that a role for NO in clot formation has been established there may be new drugs developed for the dealing with this very common occurrence.

There a number of reasons why clot formation occurs, some of them linked to the ageing process. The most common one, physical damage to the endothelium, has already been described but anything that affects the delicate balance between aggregation and disaggregation is dangerous. This can occur not only through physical damage but also from the effect of chemicals. Chemical damage may result when the production of oxygen-based radicals (such as superoxide and hydroxyl) is enhanced for some reason (such as cigarette smoking) and/or when the natural means of removing these radicals is inadequate through lack of antioxidants. When damage to the endothelium occurs, macrophages (see Chapter 11) move to the site of damage. There they ingest any oxidized lipid present to form 'foam' cells which die, leaving a lipid core coating the inside wall of the blood vessel. This is the first step in a fairly complex process called stenosis. If it continues, it can lead to serious narrowing of the lumen of an artery and restriction of blood flow. Such a lipid layer also interferes with production of NO from the endothelial cells. NO production prevents stenosis formation in a number of ways. It removes some of the oxygen-based radicals by reaction to give short-lived peroxynitrite and then nitrate (a topic dealt with in detail later) and by prevention of adhesion of the macrophages.

Stenosis most generally occurs in arteries near the heart. It is a seriously harmful condition, and an effective treatment is stenting. A stent is a tiny collapsible metal device held at the end of a catheter. Inside the stent is a balloon. The catheter is inserted into an artery in the thigh of the patient and pushed upwards until it reaches the site of the stenosis. This procedure is done with the patient awake and is followed by X-ray imaging. When the stent reaches the site of stenosis the balloon is inflated, the stent opens up and enlarges the lumen of the artery so that blood flow is no longer restricted. The catheter and balloon are then removed, leaving the stent in place (Figure 2.2). All should be well with the patient if it were not for the possibility that the insertion of the stent has further damaged the endothelium. The short-term danger (within a day of stenting) is that platelets will aggregate around the damaged endothelium because of lack of NO and form a thrombus. If this does not occur to any significant extent there is still a long-term (about three months) danger associated with a damaged endothelium. Not only does NO affect platelet aggregation and control smooth muscle relaxation, it also controls the rate at which smooth muscle cells replicate. They are dividing and growing all the time and, in healthy vessels, these processes occur in a controlled way. NO is a controlling factor for the genes that produce the enzymes responsible for cell division. In the absence of NO smooth muscle cells proliferate, resulting in a narrowing of the artery (a process called re-stenosis) and the whole point of stenting has been lost. Some patients re-stenose regularly and repeated stenting is of no value. They are best treated by means of a bypass operation. Stenting is a relatively simple procedure for treating a common condition but it is difficult to see how it could be effected without some damage to the endothelium. Coating the stent with material that slowly releases antithrombotic drugs might help. The role

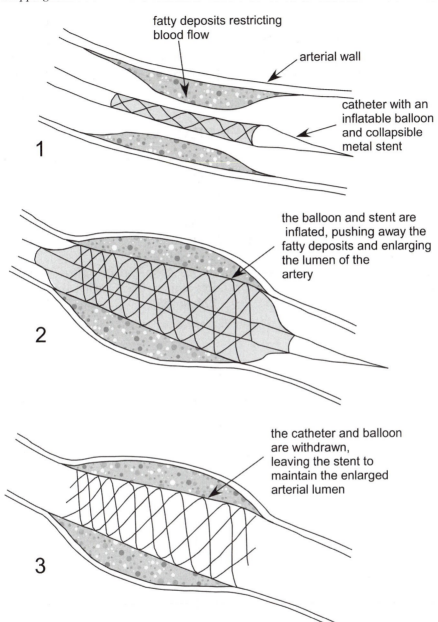

**Figure 2.2** *A stent in place in an artery*

of NO in controlling smooth muscle cell proliferation genetically was yet another unexpected discovery about NO and has consequences above and beyond the difficulties of successful stenting. NO plays a part in other genetically controlled processes, some of which are described elsewhere in this book.

## FURTHER READING

M.W. Radomski, R.M.J. Palmer and S. Moncada, Modulation of platelet aggregation by an L-arginine–NO pathway. *Trends in Pharm. Sci.*, 1991, **12**, 87.

N. Swanson, A. Stevens-Lloyd and A. Gershlick, Drug-eluting stents from lab bench to bedside. *Cardiology News*, 2001, **4**, 12.

M.S. Crane, R. Ollosson, K.P. Moore, A.G. Rossi and I.L. Megson, Novel role for low molecular weight plasma thiols in nitric oxide mediated control of platelet function. *J. Biol. Chem.*, 2002, **277**, 46858.

*Chapter 3*

# How We Make NO

NO can be formed by the combination of oxygen and nitrogen within a lightning strike and in a laboratory we can make it by the action of nitric acid on copper turnings. Neither of these routes gives any hint of how it might be made within a human cell. The route there involves an enzyme, a complex protein which takes naturally occurring substances (the substrates) and, under very mild conditions (neutral pH) and at body temperature, converts them into NO. Enzymes are responsible for almost all reactions occurring in living systems and, in general, an enzyme will catalyse only one chemical process. Enzymes are, therefore, highly selective. The enzyme (or, rather, the family of enzymes) responsible for the production of NO is known, rather unimaginatively, as the nitric oxide syntheses or NOSs. The acronym NOS is now so widely used that it has become a word in its own right. For the production of NO, NOS uses two substrates. The nitrogen of NO comes from one and the oxygen from another.

The nitrogen-providing substrate is the naturally occurring amino acid arginine, while the oxygen of NO comes from oxygen gas. It was thought initially that it might come from water. The overall reaction for the production of NO from arginine, catalysed by NOS, is shown in Scheme 3.1. The citrulline produced is recycled by other pathways in cells back to arginine.

Now that one of the substrates for NO production has been identified, one piece of evidence supporting the role of NO in blood vessels can be described. If you label one of the terminal nitrogens in arginine in a special way (by replacing the nitrogen-14 by nitrogen-15) and feed that arginine to endothelial cells, the nitrogen-15 appears in the NO formed. Only nanogram ($10^{-9}$ g) quantities of NO are produced in this way but even quantities as small as this can be detected by modern mass spectrometry. Using this specially labelled arginine the NO produced

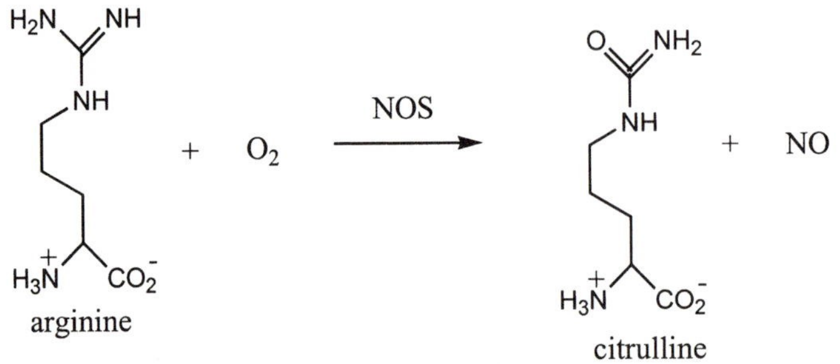

**Scheme 3.1**  *Conversion of L-arginine and oxygen into NO by the enzyme NOS*

has a relative molecular mass (what used to be called molecular weight) of 31 rather than 30, characteristic of regular NO. The detection of NO of molecular weight 31 in an early experiment by Salvador Moncada and his group was one of the first direct pieces of evidence for the production of NO from arginine in a living system.

In order to say something about the structure of the enzyme NOS we have to anticipate matters that are described in more detail later in this book: NO is found not only in blood vessels but also in macrophages, part of the body's immune system (see Chapter 11), and in nerve cells (see Chapter 15). As remarked earlier, NOS is not a single enzyme but a family of related enzymes. Members of the family may be named in one of several ways. For the sake of simplicity we will use the system where the name relates to the tissue where the NOS was first found. NOS first located in the endothelium is known as eNOS, that from macrophages as iNOS and nNOS is from brain cells. The three forms display differences in structure. The problem with this system of nomenclature is that iNOS and nNOS exist in a range of tissues only remotely related to the tissue in which they were first found. For example, iNOS exists, alongside eNOS, in vascular endothelial cells. There are two important ways in which eNOS and nNOS differ from iNOS. Firstly, eNOS and nNOS are present all the time in endothelial and neuronal tissue respectively. They may be dormant but they are there. The response to a demand for vessel dilation is very rapid, but the amount of NO produced is very small. On the other hand, iNOS is an inducible enzyme and, until it is needed, there is little of it about. When needed (following, for example, an invasion of unfriendly bacteria) the gene for iNOS is switched on to produce iNOS and large quantities of NO result, albeit produced rather slowly. Once the bacteria are destroyed the iNOS disappears. The second distinguishing feature is the response of different types of NOS to the concentration of calcium ions (Ca²⁺) within the cell. eNOS and nNOS are activated when calcium levels within a cell rise, while iNOS appears to be insensitive to calcium levels. Associated with all forms of NOS is a calcium-binding protein called calmodulin. When calcium levels in a cell rise the $Ca^{2+}/$ calmodulin complex binds strongly to eNOS and nNOS and activates the enzyme, initiating the production of NO. With iNOS a $Ca^{2+}/$ calmodulin complex is bound strongly whatever the calcium level in the cell and so iNOS appears to be insensitive to calcium levels.

We can now picture the process that occurs when acetylcholine in blood brings about NO-mediated vasodilation (Figure 3.1). On the surface of the endothelial cells are structures sensitive to acetylcholine and known as receptors. When the acetylcholine receptor is activated by acetylcholine a channel in the cell membrane opens and $Ca^{2+}$ ions flow into the cell. The raised calcium level activates eNOS to convert arginine and oxygen into NO. Some of the NO diffuses into the underlying smooth muscle cells, initiating a cascade of processes resulting in muscle relaxation and enhanced blood flow.

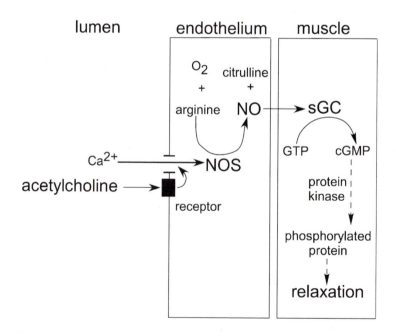

(– – – indicates sequence of reactions)

**Figure 3.1** *Activation of vascular muscle relaxation by acetylcholine with NO as the messenger molecule*

To return to the enzyme. Conversion of arginine and oxygen into NO and citrulline occurs in two stages, as shown in Scheme 3.2. The intermediate is *N*-hydroxyarginine, which can be isolated, and both stages I and II are oxygenations. The electrons necessary for these processes are supplied by an additional naturally occurring substance, the reduced form of nicotinamide adenine dinucleotide phosphate, mercifully always shortened to NADPH. NADPH is called a cofactor and, although not strictly part of the enzyme, no reaction occurs without it. When it gives up electrons it becomes NADP$^+$ and H$^+$. We can now write a balanced equation for the overall reaction (Scheme 3.3). These equations look complicated enough but they are only a fraction of the full story. NOS is an enormous enzyme and divided into two parts. One half, the oxygenase domain, receives electrons to permit the conversion of arginine and oxygen into citrulline and NO. This domain contains further cofactors: tetrahydrobiopterin (BH$_4$) and haem. These cofactors are bound strongly within the protein structure of the enzyme. Substrate oxygen binds to the iron of the haem cofactor before it oxygenates arginine to *N*-hydroxyarginine and then to citrulline, but the exact role of BH$_4$ is obscure. In the other half of the enzyme, the reductase domain, there are yet more cofactors: flavin mononucleotide (FMN)

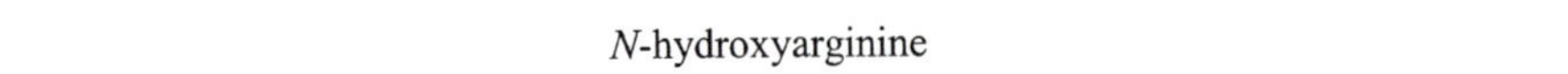

*N*-hydroxyarginine

**Scheme 3.2** *Two-stage conversion of L-arginine into NO*

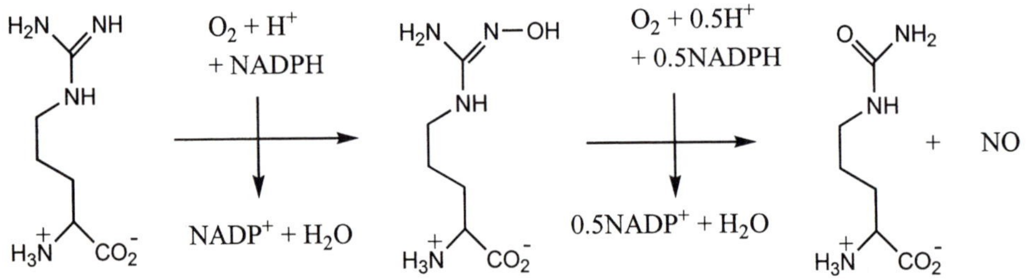

**Scheme 3.3** *Balanced equation for the conversion of L-arginine into NO*

and flavin adenine dinucleotide (FAD). FMN and FAD receive electrons from the conversion of NADPH into $NADP^+$ and $H^+$ and pass
them to the haem cofactor of the oxygenase domain, turning the oxygen
bound to its iron centre into a highly reactive form that attacks the
arginine substrate. Calmodulin/$Ca^{2+}$ appears to facilitate the passage of
electrons from FAD to FMN and it is here that the whole process is
triggered (Figure 3.2).

It has been possible to separate the oxygenase domain of human
eNOS and iNOS and determine the structures. The human iNOS oxygenase domain contains 424 amino acids and the shape has been
described as like a baseball catcher's mitt which, presumably, is similar
to that of a wicketkeeper. The haem group sits in the palm of the mitt
(Figure 3.3). There are some differences between the structures of eNOS
and iNOS and there is much interest in finding substances that will
inhibit one and not the other. This would be of value in the treatment of
septic shock, a medical condition resulting from a very general infection
and the release of much NO to counter the infection. This lowers the
blood pressure catastrophically. If all forms of NOS are inhibited there
is no NO from eNOS activity and the blood pressure goes up. In the
past such a selective inhibitor would have been sought by trial and

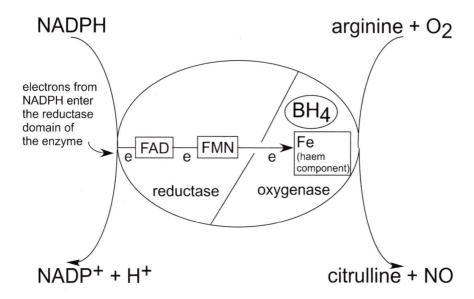

**Figure 3.2** *Diagrammatic representation of the domains of NOS*

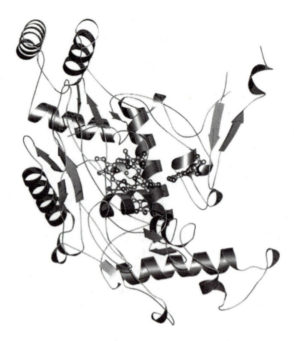

**Figure 3.3**   *Structure of the monomer of NOS with L-arginine, haem and tetrahydro-*
*biopterin at the active site; the active form of the enzyme is a dimer*

error. Now that enzyme structures can be obtained with relative ease it is possible that a selective inhibitor can be obtained by design.

In spite of an enormous amount of research on NOS (1600 research papers were published on the subject between 1989 and 2000) much remains obscure. NO is one of the smallest molecules to be biologically active and yet its formation needs one of the largest enzymes. Who or whatever is responsible for life on earth (Nature, God or evolution) obviously has a sense of humour.

The presence in blood of substances that act in the same way as acetylcholine is not the only manner in which NOS is stimulated to produce NO. The same stimulation occurs if the surrounding tissue is short of oxygen; the resulting vasodilation brings extra blood and the oxygen level is replenished. A more complex phenomenon, shear stress, has the same effect. This phenomenon is the lateral stress caused by the flow of a viscous fluid, blood in this case, over the internal surface of the blood vessel. There will be increased shear stress when the heart starts to pump more rapidly, as in a panic situation, and the stimulation of NOS results in dilation of the vessel and the arrival of more blood, required for a physical response to the panic situation.

The determination of the structure of NOS and the partial elucidation of its mode of action is one of the many triumphs of modern molecular biology. If what has been achieved leads to the synthesis of successful inhibitors then there will be ample rewards. This matter is discussed in more detail in Chapter 13.

## FURTHER READING

D. Stuehr, S. Pou and G. Rosen, Oxygen reduction by nitric-oxide synthases. *J Biol. Chem.*, 2001, **276**, 14533.

D.R. Adams, M. Brochwicz-Lewinski and A.R. Butler, Nitric oxide: physiological roles, biosynthesis and medical uses. *Fortschritte der Chemie organischer Naturstoffe*, 1999, **76**, 1.

W.K. Atherton, C.E. Cooper and R.G. Knowles, Nitric oxide synthases: structure, function and inhibition. *Biochem. J.*, 2001, **357**, 593.

*Chapter 4*

# Transporting NO

One of the most important characteristics of NO is that it is a very small molecule. Consequently it can diffuse rapidly within the vascular system, from where it has been generated (endothelial cells) to where it is required (vascular smooth muscle). In this respect it differs from many other cellular messenger molecules that are larger, some of them, such as hormones, much larger, and are transported around the body in the bloodstream. However, transport by diffusion may not be the full story, even for NO.

In the early days of NO research in biology it was generally thought that NO was produced in endothelial cells in response to local stimuli, as described in Chapter 3. For example, if a tissue becomes short of oxygen the endothelial cells of the blood vessels supplying that tissue are activated to produce more NO, causing dilation and increased blood flow. All the early experiments on NO in the vasculature used large vessels, such as aortas, because they are easier to handle. These vessels are really conduit vessels, conducting blood to different parts of the body and it is only after these vessels have divided into many, many smaller ones that there is intimate contact with tissue. To say that these smaller vessels, which are responsible for oxygenation of tissue and control blood pressure, act in exactly the same way as the larger vessels is an assumption. It has been questioned by Jonathan Stamler, a cardiologist at Duke University in North Carolina. He has proposed that NO is transported around the body to the pressure-controlling smaller vessels by attachment to haemoglobin, along with oxygen and carbon dioxide. Before this proposal can be described we have to examine the chemistry of an unusual and rather obscure class of compound: the nitrosothiols.

The thiol group, present in the important naturally occurring amino acid cysteine, is SH. In solution at body pH it is ionized to some extent to give the thiolate ion:

$$RSH \rightarrow RS^- + H^+ \tag{1}$$

Thiolate reacts readily with the positively charged species $NO^+$ to give a nitrosothiol:

$$RS^- + NO^+ \rightarrow RSNO \tag{2}$$

The compound obtained from cysteine is *S*-nitrosocysteine and it forms a bright red solution. Other thiols react in the same way. Compounds of this type were first described in the early part of the twentieth century by the Scottish chemist Alexander Macbeth (1889–1957) working at Queen's University Belfast, but they are very unstable and he was unable to characterize them fully. He little realized how important they would become. The way they decompose is interesting in that they produce NO and a disulphide (Scheme 4.1). So, *S*-nitrosothiols are formed by a heterolytic reaction (from oppositely charged species) but can decompose homolytically to give two uncharged radicals, the thiyl radical and NO. The disulphide shown below forms by the dimerization of two thiyl radicals. The process of decomposition appears to be spontaneous, and there is some spontaneous or thermal decomposition, but the main reaction is one catalysed by copper(I) ions. There is enough copper (in the order of parts per million) in good quality distilled water

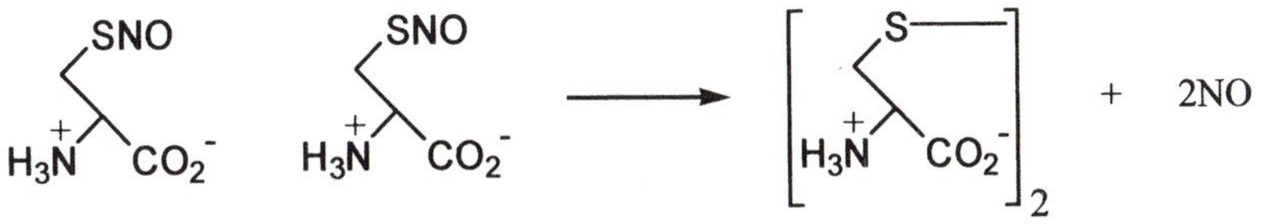

**Scheme 4.1** *Decomposition of S-nitrosocysteine to give NO*

to bring about the reaction. The naturally occurring copper in distilled water is present, of course, as copper(II) ions but copper(II) is readily reduced to copper(I) by thiols and *S*-nitrosothiols spontaneously hydrolyse to give thiols:

$$RSNO + H_2O \rightarrow RSH + HNO_2 \tag{3}$$

So, if you add an *S*-nitrosothiol to distilled water and wait a little, enough thiol is produced to convert the copper(II) to copper(I) and catalysed decomposition of the *S*-nitrosothiol commences.

There are almost certainly *S*-nitrosothiols in some cells and fluids of the body, probably not *S*-nitrosocysteine, although cysteine is ubiquitous, but *S*-nitrosoglutathione. Glutathione is a tripeptide where the central amino acid is cysteine and it exists in large amounts in many cells. *S*-Nitrosoglutathione is much more stable than *S*-nitrosocysteine and it is generally assumed to be a store of NO which cells utilize when NO is needed. Albumin, a polypeptide found in blood, also contains cysteine and there is thought to be *S*-nitrosoalbumin in blood plasma. It is difficult to say how much nitrosothiol is present in tissue as there is no reliable analytical procedure for its assay. At one time it was thought nitrosothiols were present at the micromolar level but this has been revised downwards. *S*-Nitrosothiols are biologically active, causing vasodilation and inhibiting platelet aggregation because they can release NO on demand. Whether copper(I) ions are involved is difficult to say as there is so little free copper in human tissue but any there is will probably be present as copper(I) because of the abundance of thiols. It is possible that a new class of NO-donor drugs, based on nitrosothiols, may be in clinical use in the near future.

To return to the transport of NO. Stamler has suggested that NO produced within the vasculature reacts with haemoglobin in blood and forms an *S*-nitroso compound, which acts as a vasodilator. One way in which this might occur will now be described.

Haemoglobin consists of a red, iron-containing pigment, called haem, and a protein chain. Reaction of NO with haemoglobin results in attachment of NO to the iron atom (see Chapter 9) at the centre of haem to give nitrosylhaemoglobin:

$$\text{haem} + \text{NO} \rightarrow \text{haem–Fe–NO} \qquad (4)$$

Nitrosylhaemoglobin is not a vasodilator as the NO binds very strongly to the iron. However, most NOS activity occurs in arteries, particularly in the arteries of the lung, where haemoglobin exists as its oxygenated form, *i.e.* with oxygen bound to the iron, and so the above reaction is not really relevant. The *in vitro* reaction between NO and oxyhaemoglobin has been fully explored and results in the formation of nitrate and methaemoglobin:

$$\text{oxyhaemoglobin} + \text{NO} \rightarrow \text{methaemoglobin} + \text{NO}_3^- \qquad (5)$$

In methaemoglobin the iron is present as the ferric ion ($Fe^{3+}$) while in other forms of haemoglobin it is in oxidation state 2+. The reaction between NO and oxyhaemoglobin clearly does not give a nitrosothiol but the *in vitro* experiment described above does not parallel exactly the situation in, say, the lung as the haemoglobin there may not be completely oxygenated.

The structure of haemoglobin is complex. It contains, as well as the protein chain, four haem structures with four iron atoms. If only three are oxygenated then NO, produced by endothelial cells, may bind to the iron of the fourth haem and, at some stage, transfer to an SH group of the cysteine at position 93 on the protein chain of haemoglobin, where it forms an *S*-nitrosocysteine:

$$O_2\text{–haemoglobin–Fe–NO} \rightarrow O_2\text{–haemoglobin–SNO} \qquad (6)$$

Why it should transfer in this way is not clear and the experimental evidence for this effect is complex and indirect and not accepted by all workers active in this area. Oxygenation of haemoglobin occurs at high oxygen levels, as in the lung, when the three-dimensional structure of haemoglobin is said to be in the relaxed or R state. When oxygenated haemoglobin and, possibly, $O_2$–haemoglobin–SNO suspended in blood reach capillaries where the oxygen level is low, the three-dimensional structure changes to the tense or T state and oxygen is released. At the same time, it is claimed, the T structure triggers release of NO from the SNO group. The release of oxygen as the structure of oxyhaemoglobin changes from R to T is very well explored and is not questioned, but the concomitant release of NO is a new idea. Stamler and his colleagues have produced considerable experimental evidence for this sequence of reactions but, at the time of writing, the findings have not been confirmed in other laboratories. However, the idea is a very neat one. Tissue short of oxygen receives a supply of oxyhaemoglobin at the same time as vessels in that tissue dilate through release of NO and thus increase the amount of oxygenated blood being delivered. Haemoglobin, having lost both oxygen and NO, picks up the unwanted carbon dioxide and takes it back to the lungs, where it is lost to the atmosphere. Nature appears to be working with commendable economy (Figure 4.1).

Although the levels of *S*-nitrosohaemoglobin are in some doubt it certainly exists in blood. It can be made in the laboratory and it is a vasodilator. Stamler and his colleagues have produced evidence that the concentration of *S*-nitrosohaemoglobin falls dramatically across the arterial–venous transit. That is, the concentration of *S*-nitrosohaemoglobin is high in arterial blood but declines once the blood enters the maze of small vessels from which oxygenation of tissue occurs. This is entirely consistent with *S*-nitrosohaemoglobin acting as a vasodilator in this region of the vasculature. However, not all the experimental evidence supports this view. When local production of NO in, say, the forearm is reduced by infusing an inhibitor of NOS (see Chapter 17) there is a dramatic fall in blood flow which makes sense only if the NO

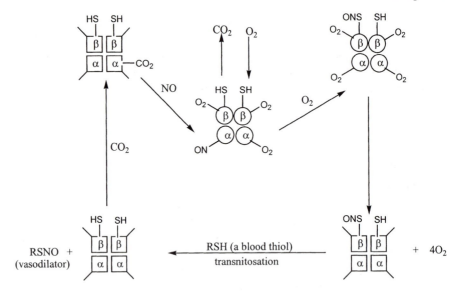

**Figure 4.1**  *Possible reactions of NO with haemaglobin during the respiratory cycle. The upper reactions occur in the lungs and the lower set in oxygen-deficient tissue. The R state is represented by circles and the T state by squares*

responsible for vasodilation is produced locally. NO transported into that tissue on haemoglobin would be unaffected by the local inhibition of NOS. Also the stimulation of NO production by shear stress (see Chapter 3) makes sense only if vasodilation is controlled by locally produced NO. It is possible, of course, that the NO required for vasodilation is both produced locally and transported attached to haemoglobin. Much more physiological research is required on this topic before the matter can be resolved. In spite of all the difficulties the transport of NO attached to haemoglobin remains an attractive idea, which, if it is true, could influence, in a major way, the production of new drugs for the treatment of cardiovascular disease.

When scientists are interviewed by the media, interviewers do not like there to be uncertainty. Media people like science to be cut and dried but, in real life, it is not like that. Imaginative ideas are proposed long before the evidence is complete. Sometimes these ideas survive the test of time, sometimes they do not. That is how science works. The transport of NO on haemoglobin would give a rapid response to the demand for oxygen. Only time, and a lot of hard work, will show if the idea is correct. There is one piece of indirect evidence to support the idea. The cysteine at position 93, on which the NO may be carried, is greatly conserved, *i.e.*, although the amino acid composition of the protein chain in the blood pigment of mammals varies considerably from one species to another, there is always a cysteine at position 93. Until now no particular use has been proposed for the conserved cysteine. Does it have the essential role of carrying NO to tissue that needs oxygen?

## FURTHER READING

D.L.H. Williams, The mechanism of nitric oxide formation from *S*-nitrosothiols (thionitrites). *Chem. Comm.*, 1996, 1085.

L. Jia, C. Bonaventura, J. Bonaventura and J.S. Stamler, *S*-nitrosohaemo-globin: a dynamic activity of blood involved in vascular control. *Nature*, 1996, **380**, 221.

A.J. Gow and J.S. Stamler, Reactions between nitric oxide and haemoglobin under physiological conditions. *Nature*, 1998, **391**, 169.

A.J. Gow, B.P. Luchsinger, J.R. Pawloski, D.J. Singel and J.S. Stamler, The oxyhaemoglobin reaction of nitric oxide. *Proc. Natl. Acad. Sci USA*, 1999, **96**, 9027.

M.T. Gladwin, J.H. Shelhamer, A.N. Schechter, M.E. Pease-Fye, M.A. Waclawiw, J.A. Panza, F.P. Ognibene and R.O. Cannon, Role of circulating nitrite and *S*-nitrosohemoglobin in the regulation of regional blood flow in humans. *Proc. Natl. Acad. Sci USA*, 2000, **97**, 11482.

M. Wolzt, R.J. MacAllister, D. Davis, M. Feelisch, S. Moncada, P. Vallance and A.J. Hobbs, Biochemical characterization of *S*-nitrosohemoglobin – mechanisms underlying synthesis, NO release, and biological activity. *J. Biol. Chem.*, 1999, **274**, 28983.

*Chapter 5*

# Drugs That Release NO

Long before NO was known to have a biological role, doctors had been using drugs which, we now know, act by releasing NO. They are effective when the body, for some reason, cannot supply enough NO of its own for dilation of blood vessels on increased demand for blood. One example clarifies the situation.

Angina is a distressing medical condition. Its symptoms are a choking or constricting chest pain that occurs behind the breastbone and may radiate to the left arm, to the throat, jaws and teeth, or to the back. It is generally brought on by physical exertion and exacerbated by emotion, large meals and cold winds. Its cause is a narrowing of the arteries supplying the heart. When increased demand is put on the heart, its tissues suffer from oxygen deprivation because of restricted blood supply. Rest can relieve the symptoms of angina, but more immediate relief comes from drugs that bring about dilation of cardiac arteries, thus enhancing the supply of blood to the heart. Such drugs have been used successfully for more than 100 years, but the reason for their success has become clear only since the role of NO in vasodilation has been known. The most common medication for angina is glyceryl trinitrate (GTN; **5.1**). Although the details are still far from clear, it is certain that GTN taken into the bloodstream is converted by enzymes into NO (Scheme 5.1). The drug is generally administered by means of a puffer but it may be taken under the tongue.

GTN has a colourful history. It is, of course, a high explosive and dynamite consists of GTN absorbed into kieselguhr, a porous material consisting largely of silica. Alfred Nobel developed dynamite around 1867 and it found widespread use in blasting. Its production was a great commercial success and gave him the fortune with which to establish the Nobel prizes, the highest honour a scientist can receive. By a curious twist of fate Nobel developed angina in his later years and was

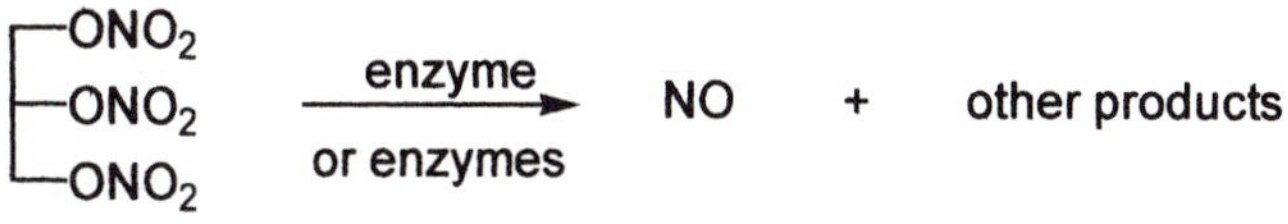

$$\begin{array}{c} \text{—ONO}_2 \\ \text{—ONO}_2 \\ \text{—ONO}_2 \end{array} \quad \xrightarrow[\text{or enzymes}]{\text{enzyme}} \quad \text{NO} \quad + \quad \text{other products}$$

**5.1**

**Scheme 5.1** In vivo *conversion of GTN into NO*

treated with the very substance that had brought him so much material success. In 1998 the Nobel Prize for Physiology or Medicine was awarded for the recognition of the role of NO in vasodilation. GTN was developed by the Italian scientist Ascanio Sobrero, who reported the first successful synthesis to the Academy of Science in Torino in 1847. Considering how dangerous a material it is (Alfred Nobel's younger brother was killed in an explosion during the development of dynamite) it is something of a miracle that Sobrero survived to report his work.

In the nineteenth century it was very common to examine the effect of newly discovered chemicals not only on animals but also on humans. Taking such risks was seen, presumably, as evidence of manliness. In 1859 a certain Mr A. G. Field of Brighton described the effects (a headache and a languid sensation) he experienced on taking a dose of a 1% solution of GTN in alcohol. Twenty years later the matter was studied more fully by William Murrell of Westminster Hospital and he identified GTN as a treatment for angina in a report published in *The Lancet*. It has been used in this way ever since.

GTN was involved in an early example of the study of environmental health. During the First World War women were employed on a large scale in factories packing GTN into munitions. Many of them complained of severe headaches and this was traced to the absorption of GTN through the skin, causing an unhealthy lowering of blood pressure. Rather curiously the headaches disappeared by the end of the week, only to return after the weekend break. Thus it was common for Swedish munitions workers to place a little GTN in the band of the hat they wore at the weekend to prevent 'Monday morning headaches'. A study in the 1950s showed that exposure to GTN caused no permanent damage to health. The disappearance of headaches after a time is a consequence of the tolerance the body manifests to GTN after a period of use and this is a major drawback in its use as a drug for the management of angina.

Other compounds related to GTN with similar clinical benefits are pentaerythrityl tetranitrate (PETN; **5.2**), isosorbide mononitrate (ISMN; **5.3**) and isosorbide dinitrate (ISDN; **5.4**) but none has achieved

5.2

5.3

5.4

5.5

the widespread use of GTN. PETN is resistant to tolerance which, as it is similar to GTN in structure, is rather difficult to understand.

Organic nitrites (such as amyl nitrite*, **5.5**) are also effective in the treatment of angina or as general hypotensive agents but are too volatile to be convenient. At one time amyl nitrite was used to treat a number of conditions other than angina. In a Sherlock Holmes story (*The Resident Patient*) Dr Percy Trevelyan, on being consulted by a client subject to cataleptic attacks, remarks "I had obtained good results in such cases by the inhalation of nitrite of amyl...". In some editions this appears as 'nitrate of amyl', a compound that would have blown not only the patient but also all the other characters in the story to eternity. During the heady days of the 1960s hippies used amyl nitrite as a recreational drug and called it 'poppers'. It probably increased the blood supply to the brain, inducing a feeling of euphoria, but did not enhance their intelligence. Both GTN and amyl nitrite are substrates for the enzyme xanthine oxidoreductase, which converts them into NO. This could be the *in vivo* process, making them effective in the treatment of angina, but very high concentrations of the two drugs are required to get the enzyme active and other routes (or other enzymes) may be more efficient.

Over the years a number of compounds containing an NO group have been used as sources of NO for the treatment of cardiovascular dysfunction. One of the most effective, but at the same time one of the most unlikely, is sodium nitroprusside, $Na_2[Fe(CN)_5NO]$. The name is unfortunate as it is not a nitro compound (which would require the presence of $NO_2$) nor is it a prusside. Its abbreviation in medical circles is SNP. Sodium nitroprusside was first made in 1848 by a Scottish

---

* This is more correctly named *iso*-amyl nitrite but amyl nitrite is the term in common usage.

chemist and, later, member of parliament Lyon Playfair (1818–1898). When infused into the bloodstream SNP is highly effective in lowering blood pressure and it is used during surgery on the vasculature. Why it is so effective is something of a mystery but the presence of NO in the anion $[Fe(CN)_5NO]^{2-}$ suggests that its action is NO-related. One possible explanation is that the intense lighting in the operating theatre releases NO from the nitroprusside ion (a known reaction) but that is certainly not the complete story. The presence of five cyano groups in SNP has led to caution in its use as death by cyanide poisoning following the use of SNP has been reported, although no figures are available. There is more about this substance as an hypotensive agent in Chapter 9.

Some new NO-liberating drugs have been developed in recent years. The *S*-nitrosothiols are described in Chapter 4 and molsidomine (**5.6**) is an example of a sydnomine that is used in the treatment of angina. In the liver it is converted into SIN-1 and this, on oxidation, releases NO (Scheme 5.2). It may also form superoxide (see Chapter 11), a considerable disadvantage in clinical use. The name sydnomine is derived from the name of the university where compounds of this type were first made, the University of Sydney.

In the 1960s Russell Drago of the University of Illinois made and characterized a number of adducts of NO with secondary amines and, for a time, these compounds were given the attractive name NONOates but the authorities changed it to the more cumbersome diazenium-diolates. Subsequent work by Larry Keefer and his colleagues at the National Cancer Institute in the USA has shown that diethylamine NONOate (DEA/NO) (Scheme 5.3) releases NO in a clean, first-order reaction with a half-life of 2.1 minutes at 37 °C and pH 7.4. Other NONOates decompose much more slowly and so these compounds provide an admirable source of NO with controlled delivery. Oxadiazoles (such as **5.7**) have been shown to display NO-related biological activity and, for some members of this family, release of NO depends upon the presence of thiols.

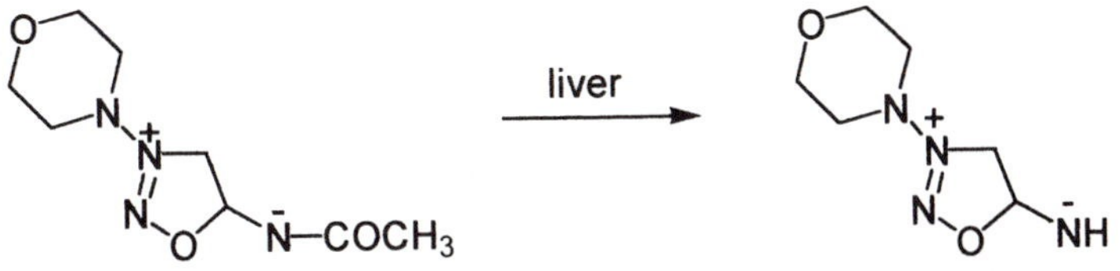

**Scheme 5.2** In vivo *activation of molsidomine*

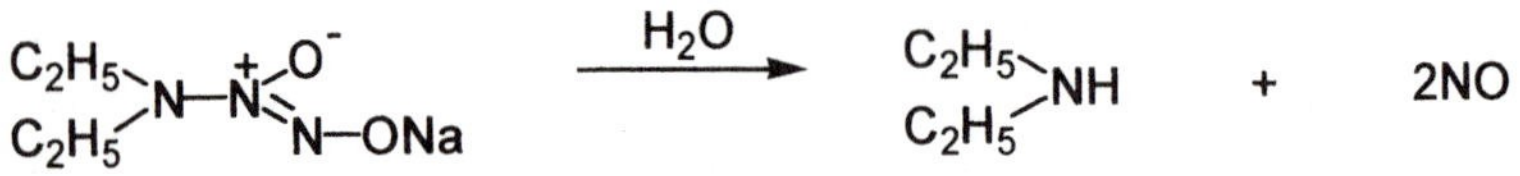

$$C_2H_5{-}N^{+}{-}N{=}N{-}ONa \quad \xrightarrow{H_2O} \quad C_2H_5{-}NH \quad + \quad 2NO$$

**Scheme 5.3**  *Production of NO from a NONOate*

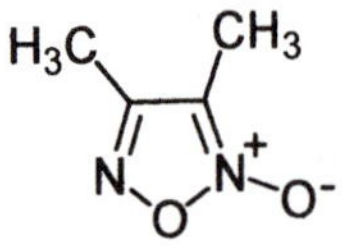

5.7

The history of medicine shows that sometimes a drug is used for treatment long before its mode of action is understood. Is there any evidence for the treatment of angina involving an NO-donor drug before the advent of modern medicine? Possibly. In 1901 an amazing collection of manuscripts were found in a walled-up cave at a Buddhist shrine called Dunhuang, in central Asia, by a British explorer, Sir Aurel Stein. They had probably been hidden there by Buddhist priests around AD 1000 as Islam swept into China. Amongst the thousands of items, everything from domestic items to religious texts in many languages, is probably the world's oldest printed book, the *Diamond Sutra*, now on display in the British Museum. The Dunhuang find includes 19 medical manuscripts and these give insight into medical practice in China around AD 800. These manuscripts have been subjected to the most rigorous scholarly scrutiny and have been translated. One describes what is obviously a treatment for the pain associated with angina (Figure 5.1). The patient is instructed to place nitre (potassium nitrate) under the tongue and leave it there while carefully conserving the saliva. This guarantees a cure. Potassium nitrate was well known to the Chinese because it is a component of gunpowder, but it is almost entirely without biological action. However, under the tongue are colonies of bacteria, some of which contain the enzyme nitrate reductase (see Chapter 18), which converts nitrate into nitrite. Nitrite can be a source of NO because of the reaction:

$$2HNO_2 \rightarrow NO_2 + NO + H_2O \tag{1}$$

Thus NO is formed from nitrite under acid conditions. Tissue starved of oxygen (ischemic tissue), which is what happens in angina, is much more acidic than healthy tissue and so if *nitrite* is taken into the

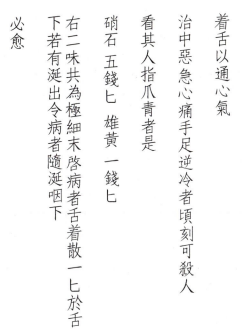

必
愈

下
若
有
涎
出
令
病
者
隨
涎
咽
下

右
二
味
共
為
極
細
末
啟
病
者
舌
着
散
一
匕
於
舌

硝
石
五
錢
匕　
雄
黃
一
錢
匕

看
其
人
指
爪
青
者
是

治
中
惡
急
心
痛
手
足
逆
冷
者
頃
刻
可
殺
人

着
舌
以
通
心
氣

**Figure 5.1** *A cure for angina in an eighth century Chinese text*

bloodstream through the tongue it could release NO in the ischemic
tissue of the cardiac arteries and bring about vasodilation. It may be
that the therapeutic use of NO in China predates its use in the West by
many centuries.

But what of the future of NO-donor drugs? It is a natural area of
research for many pharmaceutical companies. Compounds that release
NO are easy to find; compounds that do so within the relevant tissue
and nowhere else are far harder to produce, but the search goes on. One
interesting idea that has reached an advanced stage is a form of aspirin
that also releases NO. This so-called 'hybrid drug' may provide the
protective action against coronary heart disease associated with both
aspirin and NO, and possibly counteract the gastric side-effects of
aspirin.

When the blood vessels supplying the lungs have become constricted,
the idea of giving inhaled NO to dilate them, thus enhancing the blood
supply to the lungs, is an obvious one. Inhaled NO has been used in
intensive care units since 1992. However, the early work used substand-
ard gas produced for industrial purposes, together with poor delivery
systems. Inhalation of NO was, in general, a treatment of last resort,
making medical outcomes difficult to evaluate. There are considerable
dangers in the medical use of inhaled NO. On contact with oxygen

it is, of course, readily converted into toxic $NO_2$. $NO_2$ also forms by the disproportionation of NO on storage (see Chapter 8). All $NO_2$ must be removed before NO can be used clinically, but this is difficult to ensure. In recent years the use of inhaled NO for enhancing the blood supply to the lungs in adults has declined as several surveys have indicated outcomes not commensurate with the risks and difficulties involved. The sudden withdrawal of inhaled NO can also have an adverse effect on some patients.

Using inhaled NO for the treatment of newborn infants with respiratory failure has proved more successful, according to carefully controlled trials. It is generally given within 14 days of birth and reduces the need for the extreme treatment for this condition such as reoxygenation of the blood outside the body. The latter is a technique that is costly and difficult to use. In 2001 inhaled NO was registered as a drug in the European Union for the treatment of newborn infants. Why inhaled NO works better with infants than with adults is not clear. The treatment of premature babies with NO has not proved successful, possibly because the lungs are not fully developed.

## FURTHER READING

M. Feelisch and J.S. Stamler, Donors of nitrogen oxides in *Methods in Nitric Oxide Research*, M. Feelisch and J.S. Stamler (eds), Wiley, Chichester, 1996, 71.

A. Bellamy in *Atti del Convegno in celelbrazione del centenario della morte di Ascanio Sobrero*, E. Dagli (ed), Torino, 1989, 15.

W. Murrell, Nitro-glycerine as a remedy for angina pectoris. *Lancet*, 1879, 18 Jan, 80.

Z.Q. Chen, J. Zhang and J.S. Stamler, Identification of the enzymatic mechanism of nitroglycerine bioactivation. *Proc. Natl. Acad. Sci. USA*, 2002, **99**, 8306.

A.R. Butler and C. Glidewell, Recent chemical studies of sodium nitroprusside relevant to its hypotensive action. *Chem. Soc. Rev.*, 1987, **16**, 361.

G. Sorba, C. Medana, R. Fruttero, C. Cena, A. De Stilo, U. Galli and A. Gasco, Water soluble furoxan derivatives as NO prodrugs. *J. Med. Chem.*, 1997, **40**, 463; *J. Med. Chem.*, 1997, **40**, 2288.

M. Maragos, D. Morley, D.A. Wink, T.M. Dunams, J.E. Saavedra, A. Hoffman, A.A. Bove, L. Isaac, J.A. Hrabie and L.K. Keefer, Complexes of

NO with nucleophiles as agents for the controlled biological release of nitric oxide. Vasorelaxant effects. *J. Med. Chem.*, 1991, **34**, 3242.

R.H. Clark, T.J. Kueser, M.W. Walker, W.M. Southgait, J.L. Huckaby, J.A. Perez, B.J. Roy, M. Kessler and J.P. Kinsella, Low-dose nitric oxide therapy for persistent pulmonary hypertension of the newborn. *New Eng. J. Med.*, 2000, **342**, 469.

A.R. Butler and J. Moffett in *Mediaeval Chinese Medicine: The Dunhuang Medical Manuscripts*, V. Lo and C. Cullen (eds), Rutledge Curzon, in press.

*Chapter 6*

# Discovering and Making NO

It is generally stated that NO was discovered by the English natural philosopher (a gracious but outdated term for a scientist) and Unitarian minister Joseph Priestley (1733–1804), but this is not correct. As NO is readily made by the reaction of nitric acid ($HNO_3$) with any one of a number of commonly available metals (see later), the discovery of NO is not very surprising once you have nitric acid. In fact, nitric acid was known in the thirteenth century and called *aqua fortis*, and a modern way of making it (by the action of sulphuric acid on potassium nitrate) was developed by Johann Glauber (1604–1670). That a gas or 'air' was formed when nitric acid was poured over copper, iron or silver was observed by several natural philosophers, including Johannes van Helmont (1579–1644). Robert Boyle (1627–1691) and Georg Stahl (1660–1734) noted that this 'air' formed brown fumes on contact with the atmosphere:

$$2NO + O_2 \rightarrow N_2O_4 \qquad (1)$$

Stephen Hales (1677–1761), a clergyman in the then rural parish of Teddington, England, studied the formation of NO in more detail and noted that no reaction occurred on mixing concentrated nitric acid and iron filings but on addition of water NO was evolved. This observation is the origin of the phrase commonly found in older school textbooks of chemistry: 'Nitric oxide is made by the action of *moderately* dilute nitric acid on copper turnings'. There is a curious coincidence here for Stephen Hales was also the first person to measure blood pressure, not in humans but in horses. In his book *Statical Essays: Containing Hemastaticks* of 1773 he described an experiment in which he inserted a nine-foot glass tube into the artery of a horse and noted the height to which the blood rose. What Joseph Priestley did was to characterize

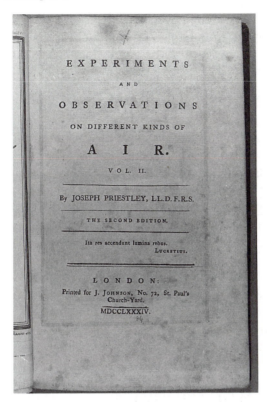

EXPERIMENTS

A N D

OBSERVATIONS

ON DIFFERENT KINDS OF

A I R.

VOL. II.

By JOSEPH PRIESTLEY, LL.D. F.R.S.

THE SECOND EDITION.

*Ita res accendunt lumina rebus.*
LUCRETIUS.

LONDON:

Printed for J. JOHNSON, No. 72, St. Paul's
Church-Yard.

MDCCLXXXIV.

**Figure 6.1**   *Title page of Priestley's book describing the properties of NO*
(Reproduced by kind permission of the National Library of Scotland)

NO as a distinct chemical entity, very different from other gases or 'airs'. The experimental work is described lucidly in his book of 1790, *Experiments and Observations on Different Kinds of Airs* (Figure 6.1). However, the impact of the work is lessened by his insistence, in spite of Antoine Lavoisier's (1743–1794) work relating oxygen and burning, on interpreting all his observations in terms of the discredited phlogiston theory.

Phlogiston was a 'principle' thought to be given off when anything burned. Priestley refers to what is obviously NO as 'nitrous air' but further confuses the situation by calling nitric acid 'nitrous acid', a name we now give to $HNO_2$. He prefaces his work on NO with the words: 'There is no kind of air, the constituent principles of which seemed to be more clearly ascertained than those of nitrous air by those philosophers who admitted the doctrine of phlogiston'. In spite of some otherwise excellent experimental science, Priestley was one of the last natural philosophers to hold on to the theory of phlogiston, even after his move to the USA because of hostility in Birmingham to his support of the French Revolution. Even so, three of his observations are worth noting. Firstly, he found that 'nitrous air' (*i.e.* NO) when stored for some time in the presence of iron and moisture diminished in volume. This presumably results from the thermodynamic instability of NO:

$$3NO \rightarrow N_2O + NO_2 \qquad (2)$$

The $NO_2$ dissolves in moisture to give nitrite and nitrate, thus diminishing the volume further. The remaining gas, what we now call nitrous oxide ($N_2O$), he referred to as 'dephlogisticated nitrous air'. Priestley's insistence that the effect is due to the presence of iron suggests that this is a catalyst for the above reaction, but the matter does not appear to be discussed in modern chemical texts. Secondly, Priestley employed NO's property of reacting with oxygen in the atmosphere to form a water-soluble gas in the volumetric analysis of air. Thirdly, he noted that NO reacts with 'green vitriol' (ferrous sulphate) to give a black solution of an iron nitrosyl, a reaction that is similar to the 'brown ring test' for nitrate so beloved of inorganic qualitative analysis in former times. Also, iron nitrosyls may have a biological role when NO acts as a cyto-toxic agent (see Chapter 11). Although his observations are impeccably reported, Priestley had no way of knowing what elements made up nitric oxide. It was the highly regarded, but also highly eccentric, scientist Henry Cavendish (1731–1810) who showed that nitric oxide contained nitrogen and oxygen, and Sir Humphry Davy (1778–1829) who proved the diatomic nature of the compound, one atom of nitrogen linked to one of oxygen.

Priestley was able to characterize NO because he had devised ways of collecting gases in gas jars by the displacement of water or, for water-soluble gases, the displacement of mercury. However, this did not allow for the ready storage of gases because gas jars leak. The problem of the storage of NO was solved in an unexpected way by a French army pharmacist called François-Zacherie Roussin (1827–1894) (Figure 6.2). He lived at a time of great social upheaval in France and spent some time in prison as he was thought to be a counter-revolutionary. In his personal laboratory he studied the chemical processes involved in the production of artificial dyestuffs. Azo dyes, which had recently been discovered, are made from the products of the action of nitrous acid ($HNO_2$) on amines, a process known as diazotization. Roussin studied the action of nitrous acid on other substances and, probably by chance, obtained an intensely black, crystalline, ionic material by mixing iron(II) sulphate, ammonium sulphide and nitrous acid. This substance has always been known as Roussin's black salt and has been considered

**Figure 6.2**   *François-Zacherie Roussin, a French army pharamacist who studied the process of diazotization*

something of a chemical curiosity. The anion was known to have the formula $[Fe_4S_3(NO)_7]^-$ but the exact structure was a mystery. In the 1970s it was subjected to analysis by X-ray crystallography and found to consist of a 'cluster' of sulphur and iron atoms to which NO groups are attached (**6.1**).

Roussin noted two interesting chemical properties of the black salt. Although ionic, it is more soluble in organic solvents than in water. Also, addition of a copper salt, acting as an oxidizing agent, to a solution of the black salt results in the evolution of very pure NO. In the laboratory, this provides a convenient way of producing the gas when required, whereas today we would use gas from a cylinder. The fact that it is soluble in lipid and can yield NO makes Roussin's black salt a good vasodilator.

**6.1**

$$C_2H_5$$
$$|$$
$$S$$

ON–Fe–NO

ON–Fe–NO

$$S$$
$$|$$
$$C_2H_5$$

**6.2**

In alkaline solution the anion of the black salt undergoes considerable rearrangement to give what has always been known as Roussin's red salt and this, on reaction with ethyl bromide, gives the incorrectly named Roussin's red ester (**6.2**). Roussin's compounds may appear to us to be rather remote from everyday modern living but this was not the case for Chinese peasants living in a remote part of Hebei Province known as Linxian. A comprehensive survey of the incidence of different types of cancer in China during the 1970s showed a very high incidence of oesophageal cancer in Linxian. A team of Chinese scientists, sent there to investigate, decided it must be a matter of diet. The most distinctive item in the peasants' food was a rather unpleasant form of pickled cabbage obtained by immersing whole cabbages in water for some days. This material was examined and found to contain small amounts of Roussin's red ester. It had never before been detected in a natural source. Some compounds containing an NO group (such as secondary nitrosamines RHN–NO) are known cancer-causing agents (see Chapter 12) and so the red ester was immediately suspected as the cause of oesophageal cancer. Its cancer-causing properties were examined and found to be slight. However, further experiments showed that Roussin's red ester is a promoter of cancerous growth that has been started in some other way and the peasants' diet was altered accordingly, with the expected beneficial outcome. It is unlikely that the peculiar set of circumstances in Linxian leading to the production of Roussin's red ester in an item of diet are duplicated elsewhere and so the ester is probably not a significant factor in global cancer causation. On the other hand NO itself does have a role in the formation of cancerous tissue and this is described further in Chapter 12.

In the years since its discovery chemists investigated the properties of NO but, apart from its role as a ligand described in Chapter 9, little of general interest was discovered. And then in January 1987 the modern NO story broke and it must now be one of the most investigated molecules of all time. That date is the start of a new era for many scientific disciplines.

Much has been written in this book about the way living systems make NO but it is also produced industrially and in laboratories. The use of NO from cylinders is not recommended for, as mentioned above, the gas is thermodynamically unstable and on storage becomes contaminated with $N_2O$ and $NO_2$. It is produced industrially only on a small scale by the catalytic oxidation of ammonia but occurs on a large scale as an intermediate in the manufacture of nitric acid from ammonia, a process known as the Ostwald process (Scheme 6.1).

Making pure NO in the laboratory is not easy. The reaction of metal with nitric acid appears to be simple and this is how the gas was discovered, but the NO obtained is rather impure. However, some of the impurities are not difficult to remove. The NO formed may be collected over water and this removes any $NO_2$ present. Care is needed in the choice of nitric acid when NO is prepared in this manner. As mentioned previously, textbooks of inorganic chemistry are careful to emphasise the word *dilute* (or, more pedantically, *moderately dilute*) when describing the nitric acid. This is because, if concentrated nitric acid is used, $NO_2$ is evolved instead of, or as well as, NO. If very concentrated nitric acid is used there is no reaction at all. Even when the nitric acid is at the right concentration to give NO, the presence of nitrous acid is required as a catalyst. Nitrous acid is usually produced spontaneously in nitric acid on storage. When more dilute nitric acid reacts with copper turnings, some $N_2O$ is also formed (Scheme 6.2). This is surprising for as poor a reducing agent as copper but is not so surprising in the reaction of zinc with dilute nitric acid. Handling NO in the laboratory needs to be carried out with care as contact with the atmosphere immediately gives $NO_2$.

For the preparation of very pure NO many methods have been proposed over the years and a brief review is given here. More details are to be found in the Further Reading list. An intriguing method published by James Ogg and Richard Ray in 1956 used previously constructed pellets which, when heated, evolve 99.8% pure NO according to the reaction:

$$3KNO_2 + KNO_3 + Cr_2O_3 \rightarrow 2K_2CrO_4 + 4NO \qquad (3)$$

Other workers have used the reaction between acidified sodium nitrite and ferrous sulphate:

$$Fe^{2+} + HNO_2 + H^+ \rightarrow Fe^{3+} + NO + H_2O \qquad (4)$$

H.L. Johnston and W.F. Giauque published an early (1929) method of producing extremely pure NO. A 50% solution of sulphuric acid is

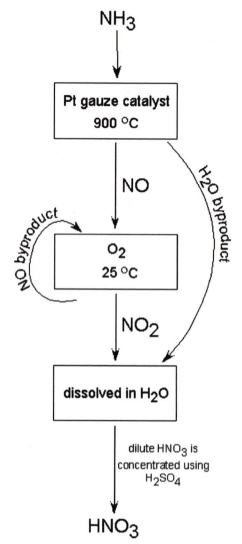

**Scheme 6.1**   *The Ostwald process for the manufacture of nitric acid*

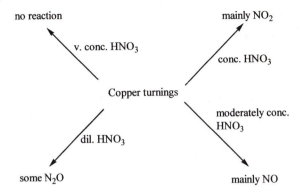

**Scheme 6.2**   *Reaction of copper with nitric acid of various concentrations*

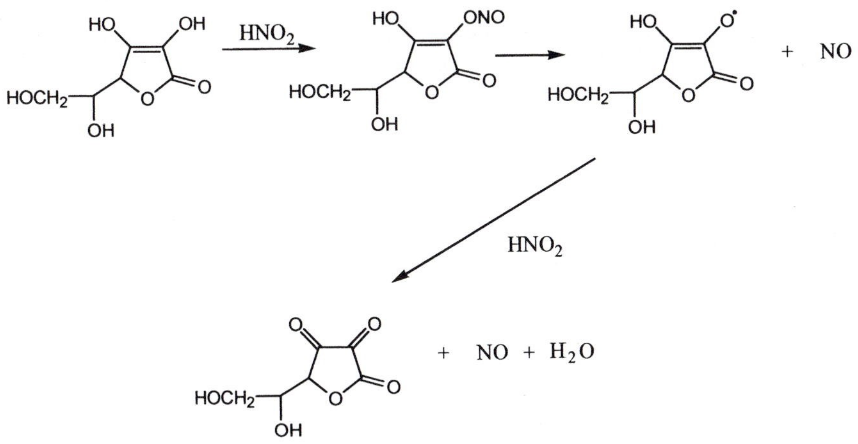

**Scheme 6.3**   *Mechanism of the reaction of ascorbic acid with nitrite*

dropped into a reaction vessel containing a mixture of KI and $NaNO_2$. NO is evolved from the solution while crystals of $I_2$ and some $K_2SO_4$ precipitate. The reaction is:

$$2NO_2^- + 4H^+ + 2I^- \rightarrow 2NO + I_2 + 2H_2O \tag{5}$$

After further purification and distillation steps, an eventual purity of 99.999% was attained.

   Probably the most convenient laboratory preparation of pure NO is the reaction between ascorbic acid and nitrite. A reasonable mechanism is *O*-nitrozation to give an ester that, by homolytic fission, gives NO and a semiquinone radical. The latter is rapidly oxidized by another molecule of nitrous acid (Scheme 6.3).

## FURTHER READING

R.A. Ogg and J.D. Ray, The quantitative oxidation of gaseous ammonia to nitrate. *J. Am. Chem. Soc.*, 1956, **78**, 5993.

H.L. Johnston and W.F. Giauque, The heat capacity of nitric oxide from 14K to the boiling point and the heat of vaporization. Vapour pressure of solid and liquid phases. The entropy from spectroscopic data. *J. Am. Chem. Soc.*, 1929, **51**, 3194.

G. Bauer, *Handbook of Preparative Inorganic Chemistry*, Academic Press, New York, 1963.

C.A. Bunton, H. Dahn and L. Loewe, Oxidation of ascorbic acid and similar reductones by nitrous acid. *Nature*, 1959, **183**, 163.

*Chapter 7*

# Making Smog – NO Becomes a Villain

Photochemical smog, for which Los Angeles became famous in the 1950s, forms when traces of volatile organic compounds are released into a warm atmosphere containing NO. A complex series of chemical reactions occurs and the energy needed for these reactions is provided by sunlight. The strong Californian sun and the abundance of cars in the Los Angeles area provide conditions ideally suited to the generation of this type of smog. However, smog is not restricted to Los Angeles and is common in cities throughout the world (Figure 7.1).

**Figure 7.1** *Sunny Fifth Avenue in a smog-covered New York City*

Most of the chemical reactions with which chemists are familiar, like the nitration of benzene or the formation of an ester, occur in solution and involve charged species. In contrast the reactions leading to a photochemical smog take place, largely, in the gaseous phase and rarely involve charged species. Instead the reactions involve radicals. A radical is a chemical species with an unpaired electron, indicated in formulae by means of an inconspicuous but vitally important superscript dot. In view of the other meanings of the word 'radical' its use as a description of a type of chemical species is unfortunate. A chemical radical is not outlandish in any way, but most radicals are highly reactive and one thing they often do is to react with another radical to give a stable molecule. NO is a radical and we should really write it as NO˙ (as biologists do) but the convention among chemists is otherwise. For reasons we have tried to explain elsewhere NO is far less reactive than many other radicals.

Only tiny amounts of NO form naturally in the atmosphere. The majority of it is produced in one of two ways: thermally or from fuel.

Thermal NO is formed by the reaction of nitrogen with oxygen. This reaction can be written simply as:

$$N_2 + O_2 \rightarrow 2NO \tag{1}$$

but the truth, as is often the case, is not that simple. The reaction occurs in a series of steps, the first of which, according to the Zeldovich mechanism proposed in 1946, is the splitting of an oxygen molecule into two oxygen atoms:

$$O_2 \xrightarrow{\text{energy}} 2O \tag{2}$$

Much energy is required for this process and it occurs during a lightning flash or at the temperature reached inside the engine of a car. The oxygen atoms formed are highly reactive and can react with a molecule of nitrogen:

$$O + N_2 \rightarrow NO + N \tag{3}$$

The process is further propagated by reaction of a nitrogen atom with another molecule of oxygen:

$$N + O_2 \rightarrow NO + O \tag{4}$$

Further oxidation of some NO gives NOx but the final amount of NOx formed is very dependent on the combustion temperature and is at a maximum around 3200 °C. However, normally only small quantities of NOx are formed in this way.

Petrol is a fairly complex mixture of substances and there are traces of nitrogen-containing compounds which, when petrol is burnt, form NO. This is known as fuel NO. The amount formed depends upon the conditions of combustion and is at a maximum under fuel-lean conditions.

Both thermal NO and fuel NO react with oxygen to give some $NO_2$. The mixtures produced by the pollution source are thus termed thermal NOx and fuel NOx. In the UK, road vehicle emissions account for over half of the total NOx emissions but the $NO_2$ in these tailpipe emissions form only a small fraction of the total emitted NOx. Most of the $NO_2$ forms after the NOx has been released into the atmosphere. More $NO_2$ could then form *via* continued reaction of NO with oxygen, but in practice this reaction is insignificant unless the concentration of NO rises beyond 500 ppm (a condition difficult to achieve even in the Los Angeles rush hour). Most of the $NO_2$ is formed *via* the fast reaction between NO and ozone:

$$NO + O_3 \rightarrow NO_2 + O_2 \qquad (5)$$

This reaction proceeds rapidly until all of the ambient $O_3$ has been used up, and yields up to 30–40 ppb of $NO_2$.

The next stage in the production of photochemical smog appears somewhat regressive and almost perverse. $NO_2$ can absorb high-energy sunlight and is broken apart to give NO and oxygen, but not ordinary molecular oxygen. Instead an oxygen atom is formed:

$$NO_2 \rightarrow NO + O \qquad (6)$$

This could react with nitrogen to give NO, but at normal temperatures this reaction does not occur, and it reacts instead with oxygen to give ozone:

$$O + O_2 \rightarrow O_3 \qquad (7)$$

Ozone causes many of the symptoms produced by smog, such as eye irritation. One of the reasons that ozone is the culprit here, rather than $NO_2$, is that ozone persists in the atmosphere far longer than $NO_2$ does. The lifetime of ozone with respect to photolysis at midday levels in Europe is 600 times that of $NO_2$. Typical $NO_2$ photolysis rates at midday are $6 \times 10^{-3}$ s$^{-1}$, whereas those for ozone are $1 \times 10^{-5}$ s$^{-1}$. This equates to average lifetimes of 3 min for $NO_2$ and 24 h for ozone. In bright sunlight, ozone splits into an oxygen molecule and an oxygen atom, but an oxygen atom in an especially excited state, known as 'singlet oxygen atom'. This is an oxygen atom in which there are no unpaired electrons, in contrast to the lowest energy state where two electrons are unpaired. It should not be confused with the term 'singlet oxygen', which is normally used to describe an oxygen *molecule* ($O_2$), in which the two normally unpaired electrons are paired or paired with anti-parallel spins. For complex reasons, in the realm of quantum physics, a singlet oxygen atom cannot recombine with an oxygen molecule but can, and does, react with water to give one of the true villains of atmospheric pollution, the hydroxyl radical:

$$O \text{ (singlet)} + H_2O \rightarrow HO^{\cdot} \tag{8}$$

Once $HO^{\cdot}$ has formed even greater complexities abound, for the hydroxyl radical is a curious species. Most radicals react with oxygen but the hydroxyl radical does not and this enhances its lifetime in an oxygen-rich environment. It does, however, react rapidly with (a) hydrocarbons (RH) and (b) aldehydes (RCHO) in the following ways:

$$\text{(a) } RH + HO^{\cdot} \rightarrow R^{\cdot} \text{ (alkyl radical)} + H_2O \tag{9}$$

$$R^{\cdot} + O_2 \rightarrow RO_2^{\cdot} \text{ (peroxyalkyl radical)} \tag{10}$$

$$\text{(b) } RCHO + HO^{\cdot} \rightarrow RCO^{\cdot} \text{ (acyl radical)} + H_2O \tag{11}$$

$$RCO^{\cdot} + O_2 \rightarrow RC(O)O_2^{\cdot} \text{ (acyl peroxy radical)} \tag{12}$$

A summary of how an alkane reacts in a photochemical smog is shown in Scheme 7.1. This mechanism is very important in smog formation, because both alkyl and acyl radicals convert NO to $NO_2$ without consuming a molecule of ozone. However, this radically formed $NO_2$ may then be photolysed and begin the cycle again as NO. Each time this happens, another molecule of ozone is generated. This is how ozone is formed photochemically and builds up in a photochemical smog.

The radicals listed in equations 9–12 above all enter further reactions with the production of more radicals by radical propagation processes. Of particular importance is the peroxyacetyl radical $CH_3C(O)-O-O^{\cdot}$. It is generated from many different hydrocarbons, aldehydes and ketones, and is important because it reacts with $NO_2$ to give peroxyacetyl nitrate, $CH_3C(O)O_2NO_2$ (PAN), another major villain in atmospheric pollution. Although it can form naturally in small amounts, PAN was first detected in photochemical smog and causes eye and respiratory irritation, is damaging to proteins and more toxic to plants than ozone.

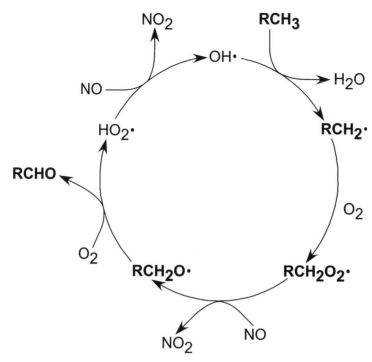

**Scheme 7.1**   *Reaction of an alkane in photochemical smog*

PAN has an unexpectedly long lifetime, 3 days at 17 °C, and much longer at lower temperatures.

When $NO_2$ forms at any stage in the complex series of reactions occurring in a photochemical smog, it can react with HO˙ to give nitric acid, which dissolves in water droplets and contributes to acid rain. Also some of the products of reaction involving hydrocarbons are themselves irritants (*e.g.* formaldehyde), adding to the harmful nature of such smog. With increasingly sensitive monitoring equipment, undoubtedly more complexities in the chemistry of atmospheric pollution will be discovered.

Photochemical smog is characterized by a daily cycle due to the time taken for the reactions to proceed. The concentration of NOx is at its peak during the morning rush hour. As the sun gets stronger towards the middle of the day, the photochemical reactions increase in number and maximum ozone concentrations are reached about 5 h after the morning rush hour. The evening rush hour also produces a peak in NOx concentration, but in the absence of sunlight to initiate the photochemical reactions, far less smog is produced.

Production of photochemical smog is not a happy situation, but once the NO has been released into the atmosphere there is little that can be done to stop it. Burning of hydrocarbon fuel will not significantly decrease in the near future, but there are ways to reduce NOx emissions. There are six strategies:

1. *Reduce the combustion temperature.* Any tactic that will reduce the temperature at the hottest part of the flame will reduce the amount of oxygen dissociated and thus the amount of thermal NOx released. Measures include injection of steam, re-injection of oxygen-depleted flue gases, fuel-rich conditions and fuel-lean conditions.
2. *Reduce the amount of time spent at peak temperature.* The flame temperature is allowed to reach a maximum, but then air, fuel or steam is injected into the burner to reduce the temperature again very quickly, so that very little oxygen dissociates.
3. *Remove the nitrogen, either by using ultra-low nitrogen fuel or by burning the fuel in oxygen instead of air.* Burning the fuel in oxygen produces an intense flame that must be diluted and dilution with air generates some thermal NOx.
4. *Sorbtion.* Injection of the effluent gas with sorbents such as carbon, aluminium oxide, limestone or ammonia can remove pollutants such NOx and sulphur. The pollutants either absorb or adsorb on to the sorbent and can later be filtered out. This is not possible for motor vehicles but can be used for large, stationary NOx sources.
5. *Chemical reduction of NOx to $N_2$.* Urea or ammonia is injected into the system to reduce the NOx on a catalyst surface. Currently this approach is not used for vehicles, only stationary combustion sources, although a similar principle is used in automobile three-way catalytic converters (see Chapter 10). In these catalytic systems, oxygen and fuel ratios are very carefully controlled so that unburned fuel acts as the reducing agent.
6. *Oxidation of NOx.* Nitrogen oxides are most soluble at higher valencies, *i.e.* $NO_3$ is more soluble than $NO_2$. This technique uses injected ozone, hydrogen peroxide or a catalyst to oxidize the NOx before it is dissolved in water. The nitric acid can be collected or neutralized to a calcium or ammonium salt and then sold.

The presence of man-made NO in the atmosphere is harmful. The small amount produced naturally from lightning, most of which ends up as nitric acid, is an important part of the natural nitrogen cycle (see Chapter 18), but the advent of the motor car and the growth of cities has created an enormous problem. Legislation and technological innovation have gone some way to solving the problem in Los Angeles but the problem is growing in other cities of the world. NO may be a benign messenger necessary for cardiovascular health but in the wrong place, like the air above an industrialized conurbation, it can be the harbinger of death.

Although the atmosphere inside city centre buildings is less polluted than that outside, the presence of pollutants indoors can still be a problem. This is particularly true for museums and art galleries. When London's National Gallery moved to its present site in Trafalgar Square in the 1830s London was a very polluted city from the burning of coal. Even then the Trustees expressed concern about the possible effect of pollution on the paintings. In 1850 the distinguished scientist Michael Faraday gave evidence to a Select Committee on the problem and recommended that the gallery should be moved westwards, where the prevailing winds would blow the pollution away. The legislation of the 1950s on burning coal in open fires brought about dramatic improvements but pollution within the Gallery, and in all city buildings holding valuable artefacts, is still a major problem. Sophisticated monitoring equipment shows just how persistent that problem is and our enhanced sensitivity to preserving the past authentically has given the solution of the problem some urgency.

Sulphur dioxide and ozone are probably the most serious pollutants as far as museums and galleries are concerned. The dangers associated with high NOx levels are more difficult to assess but, nevertheless, are real. Most of the NOx within buildings comes from the same source as that outside: the burning of fossil fuels. However, there is an added source of NOx in certain museums and that is the degradation of nitrocellulose. Where old nitrocellulose cinematographic film is stored this can give rise to high levels of NOx in the storage containers. The NOx released causes enhanced destruction of the film. Conservators also use nitrocellulose as a coating for bookbinding cloth and as an adhesive.

As far as NOx in the general atmosphere is concerned the acceptable level is set at 2–3 $\mu$g m$^{-3}$. The low level of lighting in most galleries and museums probably means that photochemical reactions of NOx are probably less important than those occurring over cities. However, the presence of NOx is not without detrimental effects. It appears that NO alone causes little damage, but its ready conversion into $NO_2$ is a source of danger to artefacts. Hydrolysis of $NO_2$ produces nitric acid (see Chapter 8), which can have a damaging effect on calcareous stone, metal objects and textiles. $NO_2$ is also an oxidizing agent and can attack polymers, increasing brittleness. As an oxidant it can react with organic pigments and dyes, affecting significant colour change. NOx probably contains some $N_2O_3$ which, as explained in Chapter 4, is a nitrosating agent. It could nitrosate the amino group of organic dyes, giving a diazonium ion:

$$(13)$$

This change from amino to hydroxyl group could have a considerable effect on the colour of a dye or pigment. $NO_2$ is said to have an effect on the arsenic pigments orpiment and realgar but it is difficult to see what the chemical process is.

Locating museums and galleries in city centres is obviously good for the user but it is the location that causes pollution problems for the conservator. Pedestrianization of city centres is a surprisingly simple solution to the problem but it is not always popular. A more sophisticated solution is to filter the air entering the display rooms, but this process is costly and requires maintenance. If our treasures are to be preserved intact for future generations the problem of pollution has to be tackled, whatever the cost.

## FURTHER READING

J.H. Seinfield, *Atmospheric Chemistry and Physics of Air Pollution*, Wiley, Chichester, 1986.

R.A. Bailey, H.M. Clarke, J.P. Ferris, S. Krause and R.L. Strong, *Chemistry of the Environment*, Academic Press, New York, 1978.

J. Heicklen, *Atmospheric Chemistry*, Academic Press, New York, 1976.

D. Saunders, Pollution and the National Gallery. *National Gallery Technical Bulletin*, 2000, **21**, 77.

P. Brimblecombe, The composition of museum atmosphere. *Atmos. Environ.*, 1990, **24B**, 1.

*Chapter 8*

# NO – a Not So Simple Little Molecule

So far this book has been concerned with what nitric oxide does rather than what it is. Why is NO, although a radical, a reasonably stable molecule? How can it react with both iron(II) and iron(III)? If NO is such a potent agent for reducing blood pressure, why is it not regularly given directly to patients? It is time to pause, and take a closer look at NO itself.

The common name for the molecule, and the one used throughout this book, is nitric oxide, but IUPAC, the body struggling to make chemical nomenclature consistent, gives it the name nitrogen monoxide. This name tells more about the molecule: *mono* indicates one, so it is made up of one nitrogen and one oxygen atom and has the chemical formula NO. NO is the second in a series of oxides of nitrogen in which the valency of the nitrogen increases, from the well-known $N_2O$ (laughing gas) to the unstable (and therefore poorly characterized) $NO_3$ or $N_2O_6$. The series is summarized in Table 8.1. Nitrogen may also access further valencies in other compounds, and for one non-metal atom to display so many ways of forming bonds with other atoms is extraordinary. There are labile reaction pathways between all these states, hence the rich chemistry exhibited by nitrogen. An excellent example is the terrestrial nitrogen cycle, discussed in Chapter 18.

Three of the oxides of nitrogen are radicals, that is to say they contain an unpaired electron in their valence shell: NO, $NO_2$ and $NO_3$. NO has eleven valence electrons, five from the nitrogen atom and six from the oxygen. A ground state molecular orbital diagram is shown in Figure 8.1. The electron occupying the highest energy level is in a $\pi^*$ (antibonding) orbital, and is unpaired. There are three electrons in antibonding orbitals and eight in bonding orbitals – this means an excess of five bonding electrons (*i.e.* 2.5 bonding pairs) and gives NO a

**Table 8.1**   *Oxides of nitrogen*

| Formula | Structure | Name | Colour | Comments |
|---|---|---|---|---|
| $N_2O$ | N≡N=O | Dinitrogen oxide (nitrous oxide) | Colourless | Rather unreactive |
| NO | N⋯O | Nitric oxide (nitrogen monoxide) | Colourless | Only moderately reactive |
| $N_2O_3$ | | Dinitrogen trioxide | Dark blue | In the gaseous phase mostly dissociated into NO and $NO_2$ |
| $NO_2$ | | Nitrogen dioxide | Brown | Reactive and harmful to humans (see Chapter 5) |
| $N_2O_4$ | | Dinitrogen tetroxide | Colourless | Partly dissociated into $NO_2$ as a liquid and mostly associated as a gas |
| $N_2O_5$ | | Dinitrogen pentoxide | Colourless | Forms an ionic solid Unstable as a gas |
| $NO_3$, $N_2O_6$ | | Nitrogen trioxide | | Unstable and poorly characterized, but known as a reaction intermediate |

formal bond order of 2.5. Certain physical properties of NO reflect this. The bond length in NO is 1.15 Å, compared with the triple bond length of 1.06 Å in $NO^+$ and a typical double bond length of about 1.2 Å. Likewise the infra-red stretching frequency of NO is 1877 cm$^{-1}$, compared with 2200 cm$^{-1}$ for $NO^+$. Other physical properties of NO are listed in Table 8.2.

Back to that single electron in the antibonding π-orbital. Any chemical species with an unpaired electron is a radical. General discussions involving radicals are usually concerned with their reactions because the majority of radicals are so reactive that further reaction occurs rapidly once the free radical has formed. Why doesn't this happen with nitric oxide? For example, NO might be expected to dimerize:

$$2NO \rightarrow ON–NO \tag{1}$$

This would eliminate the unpaired electrons by using them to form a nitrogen–nitrogen bond but there is a loss of entropy in this process and so a lower energy state is not reached. For this reason the reaction does not occur under normal circumstances at room temperature. However, in the solid state, X-ray studies have shown that NO does exist as a dimer but the molecules are joined side-by-side rather than end-to-end (**8.1**).

Investigation by magnetic methods has shown that at its melting point liquid NO is 97% dimeric, at its boiling point it is 95% dimeric,

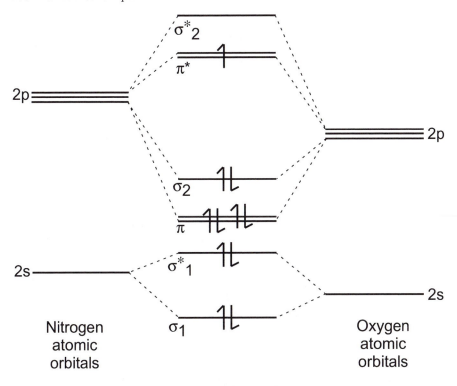

**Figure 8.1**   *Ground state molecular orbital diagram for NO*

**Table 8.2**   *Physical properties of NO*

| | |
|---|---|
| First ionization potential of NO | 9.25 eV ±0.02 (cf. CO 14.1 eV, $N_2$ 14.5 eV) |
| Infra-red stretching frequency vNO | 1877 cm$^{-1}$ (cf $NO^+$ vNO = 2200 cm$^{-1}$) |
| Melting point | −163.6 °C (109.49K) |
| Boiling point | −151.8 °C (121.36K) |
| $\Delta H$ vapourization at 121.36K | 3.292 kcal mol$^{-1}$ |
| $\Delta H$ fusion at 0K | 21.45 kcal mol$^{-1}$ |
| $\Delta H_f$ (enthalpy of formation) at 298K | 21.57 kcal mol$^{-1}$ |
| $\Delta G_f$ (Gibbs energy of formation) at 298K | 20.69 kcal mol$^{-1}$ |
| $C_p^\circ$ (heat capacity) at 298K | 7.133 cal deg$^{-1}$ mol$^{-1}$ |
| Density of solid | 1.57 g cm$^{-3}$ at 20K* |
| | 1.556 g cm$^{-3}$ at 78K* |
| | 1.46 g cm$^{-3}$ at 98K* |
| Density of gas | 1.3402 g l$^{-1}$ at 293K |

\* Calculated using X-ray techniques

$$N \overset{2.18}{\text{----}} N$$

101° 1.15

$$O \overset{\text{------}}{\underset{2.62}{}} O$$

**8.1**

**8.2**

N—O

**8.3**

but that the gas is more than 99% monomeric. The small amount of dimer that persists in the gaseous phase has a longer N–N distance (2.237 Å) and a smaller O–N–N bond angle (99.6°) than the solid state dimeric structure.

This lack of dimerization is one example of the stability of NO compared with many common carbon-centred radicals. In carbon-centred radicals, such as the methyl radical, $CH_3 \cdot$, the unpaired electron is in a *p*-type orbital centred on the carbon atom (**8.2**). It is possible to stabilize a carbon-centred radical by introducing substituents that allow interaction between the *p*-type singly occupied molecular orbital (SOMO) and the lowest unoccupied molecular orbital (LUMO). The LUMO is delocalized over the central carbon atom and the surrounding atoms and the resulting delocalization of the single electron stabilizes the molecule. In the same way, the single electron in NO is not centred on the nitrogen atom; it is in a $\pi^*$ orbital that is delocalized between the nitrogen and oxygen atoms (**8.3**). Experiments using magnetic studies have shown that the single electron is only near the nitrogen atom about 60% of the time. Delocalization of the single electron is an important factor in accounting for the lack of reactivity of NO.

What happens to NO if you leave it alone? The above discussion about the stability of NO might give the impression that a gas jar full of NO would, if left to stand, do nothing, but that is not the case. A reaction becomes possible if the energy of the products is lower than that of the reactants and the energy released as the reaction proceeds generally appears as heat. Occasionally students are advised to carry out a reaction with a supply of ice on hand to cool the reaction vessel if the temperature begins to rise out of control. This is thermodynamics in action and a simple example of First Law of Thermodynamics. The Gibbs free energy function is a means of measuring the energy difference between products and reactants. Now examine the Gibbs free energy function for the decomposition or disproportionation reactions of NO shown in equations 2 to 5. In each case, the change in energy is negative, *i.e.* the energy of the products is lower than that of the reactants and the reactions would therefore be expected to proceed. From these data it might be inferred that if a sealed vessel containing

$$2NO(g) \rightarrow N_2(g) + O_2(g) \qquad \Delta G^\ominus = -41.3 \text{ kcal} \qquad (2)$$

$$4NO(g) \rightarrow N_2(g) + 2NO_2(g) \qquad \Delta G^\ominus = -57.8 \text{ kcal} \qquad (3)$$

$$4NO(g) \rightarrow O_2(g) + 2N_2O(g) \qquad \Delta G^\ominus = -32.7 \text{ kcal} \qquad (4)$$

$$3NO(g) \rightarrow N_2O(g) + NO_2(g) \qquad \Delta G^\ominus = -24.6 \text{ kcal} \qquad (5)$$

pure NO were left to stand for long enough, a mixture of $N_2$, $O_2$, $N_2O$ and $NO_2$ would be found when it was reopened. Why, then, is it possible to purchase a cylinder of NO? Does something prevent these reactions from taking place? Well, yes and no. What thermodynamics does not tell us is the size of the activation energy of any particular reaction and it is the activation energy that fixes how fast a reaction occurs. For reactions 2–5 the activation energies are high and the rates of reaction are so slow that the NO in a cylinder is still largely unchanged, even after some weeks of storage, but it will be contaminated with some $N_2$, $O_2$, $N_2O$ and $NO_2$. At high pressures the reactions are faster. The presence of contaminants is more than just a nuisance to chemists working with NO, because $NO_2$ can initiate the explosive decomposition of hydrocarbons such as the waxes and greases found on vacuum lines. It is also toxic to humans, limiting the use of NO as a medical inhalant (Chapter 5).

The overall reaction for the disproportionation of NO is shown in equation 5. Experimentally the rate of this reaction is dependent on the concentration of NO to the power of 3. In other words, the speed at which NO disappears is proportional to the concentration of NO cubed. In chemical notation this is written:

$$\frac{-d[NO]}{dt} = k'[NO]^3 \qquad (6)$$

This is an example of a third-order reaction and it suggests that the mechanism of the reaction is not as simple as it appears because the simultaneous collision of three molecules is a very rare event. Another unusual feature of this reaction concerns the rate constant, $k'$ in the rate equation above. The rate constant for the reaction would be expected to increase with increasing temperature, *i.e.* the reaction would be expected to proceed more quickly as energy is added to the system and the temperature is raised. But for this reaction the rate constant hardly varies when the reaction temperature is doubled. The reason for this is found in the mechanism of the reaction. It is widely accepted that the reaction begins with the rapid equilibrium dimerization of a very small amount of NO:

$$2NO(g) \overset{k}{\rightleftharpoons} N_2O_2(g) \tag{7}$$

The $N_2O_2$ then reacts further with NO:

$$N_2O_2(g) + NO(g) \overset{k}{\rightarrow} N_2O(g) + NO_2(g) \tag{8}$$

So another way of writing the rate equation (equation 6) is:

$$\begin{aligned} \text{rate} &= k'[N_2O_2][NO] \\ &= Kk[NO]^3 \end{aligned} \tag{9}$$

where $K$ is the equilibrium constant for equation 7 and $k$ is the rate constant for equation 8. This explains how we can have a third-order reaction without the statistically unlikely three-molecule collision. The fact that the rate of the reaction hardly varies with temperature may be explained by noting that although $k$ increases with temperature, the value of $K$ decreases and the product $kK$ remains essentially unchanged.

The reactions described above take place largely in the gas phase, but the body is 70% water and biological NO is in solution. In order to begin to understand how NO can act as a signalling agent in this aqueous environment, we must consider how NO behaves in water. Some other oxides of nitrogen react with water to form acids, *i.e.* they are anhydrides of acids. $N_2O_5$ is the anhydride of nitric acid (equation 10) and $N_2O_3$ is the anhydride of the much weaker nitrous acid (equation 11):

$$N_2O_5 + H_2O \rightarrow 2H^+ + 2NO_3^- \tag{10}$$

$$N_2O_3 + H_2O \rightarrow 2HNO_2 \tag{11}$$

It was once thought that NO was the anhydride of the weak acid $H_2N_2O_3$, because the latter decomposes to give NO and $H_2O$:

$$H_2N_2O_3 \rightarrow 2NO + H_2O \tag{12}$$

but $H_2N_2O_3$ decomposes *via* an irreversible and complicated chain process and there is no acid–anhydride relationship between NO and $H_2N_2O_3$. Another candidate for an acid–anhydride relationship with NO is the rather rare hydronitrous acid ($H_2NO_2$) because it could theoretically be formed by the reaction of NO with water. However, isotope exchange studies have demonstrated that NO undergoes no hydrolysis at all in an aqueous environment. NO may indeed be described as the formal anhydride of hydronitrous acid, but the description is rather

meaningless. The lack of interaction between NO and $H_2O$ is essential
for the function of NO in the body. It can be generated in one cell,
released into the aqueous environment between cells, and diffuse
*unaltered* until it reaches a cell membrane provided that there is little
oxygen around. Cell membranes are constructed of non-polar lipids, so
the small, hydrophobic NO molecules can diffuse straight through and
interact with enzymes in the new cell.

The description of NO as hydrophobic suggests that it is only spar-
ingly soluble in water. This comes as no surprise since the NO molecule
does not have much of a dipole moment, which is necessary for solution
in a polar substrate such as water. Its solubility is approximately
$1.7 \times 10^{-3}$ mol $l^{-1}$ at 25 °C and 1 atm. This is similar to other non-polar
diatomic molecules such as $O_2$, $N_2$ and CO. The solubility of $N_2O$,
which has a small dipole, is 10 times greater.

It is of particular interest that the solubility of NO in water is similar
to that of $O_2$ as the best-known reaction of NO is that with $O_2$. When a
gas jar of NO is opened to the air, brown fumes of $NO_2$ are formed
immediately. However, in solution the reaction takes a different course.
$NO_2$ is still formed by the reaction of NO with oxygen, but then the
$NO_2$ rapidly reacts with another molecule of NO to form the anhydride
of nitrous acid:

$$NO_2 + NO \rightarrow N_2O_3 \tag{13}$$

$$N_2O_3 + H_2O \rightarrow 2HNO_2 \tag{14}$$

The reaction of $NO_2$ with NO is, in aqueous solution, faster than the
hydrolysis of its dimer and so NO is quantitatively converted into
nitrite. However, nitrite is readily oxidized to nitrate and so the nitrite
may be contaminated with nitrate. In some early experiments on the
behaviour of NO in aqueous solution, oxygen was not rigorously
excluded and what were thought to be the reactions of NO were those
of $NO_2$ and $N_2O_3$.

When NO is present in an aqueous biological environment there is
an additional complication. The rate of reaction between oxygen and
NO *in vitro* is such that it would appear that the oxidation of NO is not
significant *in vivo* because of the very low concentrations in living
tissue of both species. However, the cellular environment is not pure
water and there is good experimental evidence that in a heterogeneous
environment containing both water and lipid, oxidation to $N_2O_3$ is
much accelerated. This observation has two consequences. Firstly, it
is sensible to use nitrite levels in tissue as measures of NO activity

and, secondly, nitrosation of thiols by $N_2O_3$ can and will occur. The significance of the latter will be clear later.

It has been suggested that part of the versatility of NO lies in its being both easily oxidized and easily reduced. NO has a low ionization potential. The eleventh electron in the valence shell of NO is not strongly held by the molecule because it is in an antibonding orbital. Removal of this electron gives the nitrosonium ion, $NO^+$, which is isoelectronic with CO and, like CO, contains a strong triple bond. The nitrosonium ion forms two main types of compound. The first is ionic salts such as $NO^+BF_4^-$. In water, these salts behave very differently from NO. They are rapidly hydrolysed, forming nitrous acid:

$$NO^+ + H_2O \rightarrow H^+ + HNO_2 \tag{15}$$

$NO^+$ can be detected in aqueous solution only at very low pH. This means that in the body, $NO^+$, if produced, would only have a transient existence. However, there are instances of nitrosation reactions, where the $NO^+$ moiety is transferred from one molecule to another. An important example is the transfer of an $NO^+$ group from a nitrosothiol to a second thiol or another nucleophile:

$$RSNO + R'S^- \rightarrow RS^- + R'SNO \tag{16}$$

Such reactions are thought to be biologically important. The second type of compound found is one in which there is a covalent bond to give compounds such as NOX. Good examples of the latter are the nitrosyl halides:

$$2NO + X_2 \rightarrow 2XNO \ (X = F, Cl, Br) \tag{17}$$

The nitrosyl halides hydrolyse rapidly in water:

$$XNO + H_2O \rightarrow H^+ + X^- + HNO_2 \tag{18}$$

NO may be reduced by a number of reagents, and the degree of reduction may be more or less drastic. Products range from the five-electron change to ammonia to the one-electron change to $N_2O$ or $N_2O_2^{2-}$. Simple one-electron reduction of NO gives the nitroxide ion, $NO^-$. This ion is isoelectronic with $O_2$ and, like $O_2$, has two electrons in antibonding orbitals with parallel spins in the ground state, a rare example (along with oxygen) of a triplet ground state. The first excited state is a singlet and the triplet/singlet gap is 17–21 kcal mol$^{-1}$. The nitroxide ion is said to be produced chemically *via* decomposition of

$HN_2O_3^-$ or Piloty's acid ($C_6H_5SO_2NHO^-$) and ionization of the resulting nitroxyl (HNO):

$$HN_2O_3^- \rightarrow HNO + NO_2^- \tag{19}$$

$$C_6H_5SO_2NHO^- \rightarrow C_6H_5SO_2^- + HNO \tag{20}$$

$$HNO \rightarrow H^+ + NO^- \tag{21}$$

Even *in vitro* the nitroxide ion is rarely observed because HNO undergoes dimerization to give $N_2O$ and water:

$$2HNO \rightarrow N_2O + H_2O \tag{22}$$

The solution chemistry of nitroxide has been extensively investigated following formation by pulse radiolysis.

There has been much speculation about a biological role for the nitroxide anion. Much depends on the possibility of its being produced by biological reduction of NO. The widely reported value of +0.39 V for the reduction potential of $NO/^3NO^-$ made the *in vivo* production of $NO^-$ possible. However, a reduction potential of –0.8 V has been obtained by the use of very dependable procedures and it appears that $NO^-$ cannot be produced by biological reducing agents that can convert $O_2$ into superoxide. This value of the reduction potential necessitates a marked revision of the $pK_a$ of HNO from 4.7 to 11.6 (±3.4). This means that at physiological pH $HNO/NO^-$ exists exclusively as the protonated form. Therefore a biological role for $NO^-$ is not possible although it can participate in reactions that appear to be biologically significant, such as binding to the iron in haemoglobin and modifying thiols. At one time it was thought to have a biological role and, indeed, there was a theory that NOS first produced $NO^-$ and this was later oxidized to NO. At the time of writing there is little interest in this idea and a biological role for $NO^-$ now seems less likely. It is worth reflecting that the dimerization of HNO is a second-order reaction with respect to HNO concentration so at the very low concentrations that might be found in cells dimerization would be slow.

Now we can begin to see why this tiny little molecule called nitric oxide can have such a huge effect in places as diverse as the atmosphere and our bodies. It shares some of the characteristics that make it an ideal messenger molecule, such as its lack of reactivity with water, with other similar species such as $O_2$ and CO. However, in other ways, such as its bonding to transition metals and redox chemistry, it is singular. The next chapter examines the interaction between NO and transition metals, a group of reactions crucial to the action of NO in the body.

# FURTHER READING

F.T. Bonner and G. Stedman, The Chemistry of Nitric Oxide and Redox-related Species *in Methods in Nitric Oxide Research*, M.Feclish and J.S. Sramler (eds), John Wiley & Sons, 1996, Chapter 1.

T.P. Melia, Decomposition of nitric oxide at elevated pressures, *J. Inorg. Nucl. Chem.*, 1965, **27**, 95.

H.H. Awad and D.M. Stanbury, Autoxidation of nitric oxide in aqueous solution. *Int. J. Chem. Kinetics*, 1993, **25**, 375.

X. Liu, M.J.S Miller, M.S. Joshi, D.D. Thomas and J.R. Lancaster, Accelerated reaction of nitric oxide with $O_2$ within the hydrophobic interior of biological membranes. *Proc. Natl. Acad. Sci. USA*, 1998, **95**, 2175.

A.R. Butler, F.W. Flitney and D.L.H. Williams, NO, nitrosonium ions, nitroxide ions, nitrosothiols and iron-nitrosyls in biology, *Trends in Pharmacolog. Sci.*, 1995, **16**, 18.

The oxides of nitrogen are covered in many inorganic textbooks. See, for example, F.A. Cotton and G. Wilkinson, *Advanced Inorganic Chemistry*, 6th edn, Wiley, Chichester, 1999.

# Chapter 9

# NO and Transition Metals

NO produced in an organism acts as a signalling agent by activating a number of different enzymes. The enzymes all have one thing in common: a metal, usually iron but possibly copper or zinc. The reactions of NO with transition metals have been studied for many years. Indeed, many students of chemistry will have made a transition metal complex containing a NO (nitrosyl) ligand without perhaps realizing it. The brown ring in the classic laboratory test for nitrates is none other than the iron complex $[Fe(H_2O)_5NO]^{2+}$. A number of metal–NO complexes have been made and extensively studied, partly because the industrial application of such complexes, in areas such as catalysis, was realized early on. Now these studies greatly enhance our understanding of the way in which NO acts biologically.

There have been many debates over how exactly the NO binds to the metal for early attempts at describing the bonding relied on formal electron-counting methods. Such techniques are rather unhelpful in this case. It is better to think in terms of which electron orbitals on the metal interact with which orbitals on the NO. NO is an example of a so-called π-acid ligand. The archetypal π-acid ligand is carbon monoxide (CO), so it is valuable first to look at the bonding of CO to metals. When CO binds to a metal centre, a σ bond (like that between carbon and hydrogen in an organic molecule) is formed *via* overlap between a filled non-bonding orbital containing the lone pair of electrons on CO and an empty σ-type orbital on the metal. However, in the case of many metals, and especially those in oxidation state of +2 or lower, this puts excess electron density at the metal centre. The metal responds by pushing electron density away from its filled $d_{xy}$, $d_{xz}$ and $d_{yz}$ orbitals to overlapping, empty π* orbitals on the CO (Figure 9.1). These orbitals overlap because the CO is positioned end-on to the metal centre, *i.e.* the metal, carbon and oxygen atoms are in a straight line. The ability of

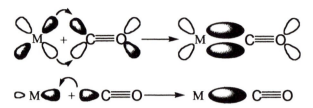

**Figure 9.1**   *Formation of a synergic bond between CO and a metal. A σ-type bond is*
                *formed via overlap of an s-type orbital on the metal and a non-bonding orbital*
                *on the CO (bottom), whilst a p-type bond is formed via overlap of a d orbital*
                *on the metal centre and a π\* antibonding orbital on the CO. For clarity, only*
                *one of the two orthogonal bonds is illustrated*

the CO ligand to accept electron density back from the metal *via*
π-bonding is called π-acidity. Overall this kind of bonding is called
synergic, because as more electron density is donated to the metal *via*
the σ bond, more can be back-donated to the ligand *via* the π bond.
Both processes strengthen the metal–carbon bond.

NO has just one more electron than CO, and that electron makes
quite a difference. As described in Chapter 8, the extra electron is in one
of the π\* orbitals to which the metal will back-donate electron density.
On the surface therefore it would appear that NO would be less able to
accept π back-donation, but now we come to the unique properties of
NO. NO can lose or gain electron density relatively easily, and do so as
the bond to a metal is being formed, giving to or taking from the metal.
NO can co-ordinate to a metal in any state in between the two extremes
of $NO^+$ and $NO^-$. Exactly what happens depends on the metal, the
oxidation state of the metal, the other ligands in the complex and
presumably environmental factors such as temperature and solvent.
Attempts have been made to predict which mode of bonding will occur
in a particular complex, but for every rule there are exceptions; as yet
the situation is just not well enough understood. It is commonplace to
visualize the bonding by first considering the cases of $NO^+$ and $NO^-$,
but remember that in a real metal–nitrosyl complex the metal–nitrogen
bond will be somewhere in between these two extremes.

$NO^+$ is isoelectronic with CO (it has the same number of electrons)
therefore if the ligand were formally $NO^+$ the bonding would be similar
to that in a metal carbonyl. As in a metal carbonyl, the $NO^+$ ligand is
bonded linearly (that is to say the metal, nitrogen and oxygen are in a
straight line), and the resulting synergic bond between the metal and
NO is strong. Now it is possible to see how one can begin with either
NO gas or acidified sodium nitrite (a source of $NO^+$) and finish with a
complex that looks the same, although the latter complex will have one
fewer electron in total.

The nitroprusside ion, $[Fe(CN)_5NO]^{2+}$, mentioned in Chapter 5, is an example of a complex where NO is co-ordinated linearly. Prior to the 1960s it was generally assumed that NO bound to a metal would be bonded linearly, despite the doubters who proposed that alternatives should be considered. There was just no evidence available at the time that proved NO was not co-ordinated linearly in a known complex. Then, in 1967, James Ibers and Derek Hodgson at Northwestern University described a complex in which the NO ligand was bent with respect to the metal, $IrCl(CO)(NO)(PPh_3)_2$ (**9.1**). This led to consideration of the other extreme, namely $NO^-$ bound to the metal centre.

$NO^-$ contains half-full $\pi^*$ orbitals, so the possibilities for back-bonding become rather restricted. Instead, the bond to the metal is formed *via* $\sigma$ donation only from a $sp^2$ hybrid orbital. The shape of the $sp^2$ hybrid orbital causes the $NO^-$ ligand to become bent with respect to the metal centre, at an angle of about 120° (Figure 9.2). Despite the lack of back-donation, the bond formed to the metal is still very strong. $NO^-$ is isoelectronic with $O_2$ and oxygen usually binds to metals in exactly the same way as $NO^-$ (as shown in Figure 9.2), for example, as in oxyhaemoglobin.

In all complexes where NO is bound to one metal centre, the bonding is thought of as being somewhere in between these two extremes. In complexes containing linearly bonded NO, the NO is almost completely bound as $NO^+$, whilst in complexes where the NO is bent at an angle of 120° it is almost completely $NO^-$. It is of course possible to have partial $sp^2$ hybridization of a lone pair in the nitrogen atom combined with partial $d-\pi^*$ overlap, for example the complex $[RuCl(NO)_2(PPh_3)_2]^+$ (**9.2**). This complex is square pyramidal, with an angularly bonded

$$\begin{array}{c} O \\ \backslash \\ N \hspace{0.3em} 124° \\ | \\ Cl\cdots\cdots\hspace{-0.5em}Ir\hspace{-0.5em}\cdots\cdots PPh_3 \\ Ph_3P \hspace{1em} CO \end{array}$$

**9.1**

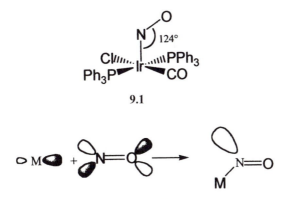

**Figure 9.2** *Formation of a σ-type bond via overlap between an s-type orbital on the metal and an sp² hybrid orbital on the NO. Note that the M–N–O bond angle is 120°*

9.2

NO in the apical (top) position and a linearly bonded NO in the square plane. The bent NO ligand has an Ru–N–O bond angle of 136°. A few similar cases are known, such as $MoCl_2(NO)_2(PPH_3)_2$ (in which the mean Mo–N–O angle is 162°) and $Co(S_2CNMe_2)_2NO$ (here the NO ligand has two orientations with a mean angle of 135°).

There are two further bonding modes that must be considered here: side-on and bridging. $NO^-$ is isoelectronic with $O_2$, and, in addition to the bonding mode similar to that shown in Figure 9.2, $O_2$ can some-times bind to metals in a side-on fashion so that both oxygen atoms are the same distance from the metal. Although theoretical calculations have suggested that, for systems with very few electrons, side-on bond-ing may be preferred, it has been reported in only a meta-stable nitrosyl complex generated photochemically in a low-temperature solid matrix. The main reason for this is probably associated with the small per-manent dipole moment of NO, which ensures that the nitrogen atom is always slightly more electron-rich than the oxygen and hence it feels a greater attraction to the positive metal. This tips the balance in favour of bonding only through the nitrogen atom. Similar considerations explain why CO binds to metals only through the carbon atom and never the oxygen.

NO as a bridging ligand between two metal centres has been observed in some cases. In the complex $Pt_4(MeCO_2)_6(NO)_2$, the metal–nitrogen–metal bond angle is 120°, and the bridging is similar to that observed in many carbonyl complexes. One way of considering the bonding is as shown in Figure 9.2, but with the lone pair on nitrogen donating to another metal atom. This is then analogous to bridging by Cl.

The ways in which NO binds to transition metals have some very important implications. For instance, metals containing CO as a ligand usually undergo displacement of the CO via a dissociative pathway, as the CO leaves the metal first. However, in many cases where a metal is bound to both CO and NO, the CO is replaced *via* an associative path-way (the incoming ligand co-ordinates to the metal centre before the CO leaves). Presumably this is because NO co-ordinated linearly as $NO^+$ can temporarily withdraw electron density from the metal centre and revert to bent $NO^-$-type bonding in which the metal has two fewer electrons, enabling it to accommodate an incoming ligand. Since the total number of electrons in the outer shell of the metal is two more for linear than for bent NO, this change in bonding would make a co-ordination site for the incoming ligand available. The ability of an NO ligand to act as an electron reservoir during a reaction may facilitate reactions that would not occur without the NO ligand. In addition, NO co-ordinated to a metal centre may be prone to reactions that do not occur with free NO. If NO is co-ordinated to a metal centre as $NO^+$, and there is strong $\sigma$ donation but weak $\pi$ back-donation, the nitrogen atom may become vulnerable to nucleophilic attack. Such a reaction occurs with the nitroprusside ion, $[Fe(CN)_5NO]^{2-}$. The CN ligands are strongly electron withdrawing, so electron density is pulled towards the metal in the metal–nitrogen $\sigma$ bond. Similarly, $\pi$ back-bonding is weak, and $[Fe(CN)_5NO]^{2-}$ reacts with thiols ($RS^-$) to give a red complex, a classic test for the presence of $RS^-$:

$$[Fe(CN)_5NO]^{2-} + SR^- \longrightarrow \left[ Fe(CN)_5N \diagup\!\!\!\!\!\diagdown {}^O_{SR} \right]^{3-} \qquad (1)$$

The reaction of $[Fe(CN)_5NO]^{2-}$ with thiols is undoubtedly part of the explanation of nitroprusside's hypotensive action, as described in Chapter 5. Direct reaction of co-ordinated NO with oxygen, usually in the presence of a base, has also been observed, but the mechanism of this reaction is still a matter for discussion.

In $NO^+$-type complexes where $\pi$ back-bonding to the nitrosyl ligand is strong, electrophilic attack by $H^+$ has been observed on the N, the O and both. Such reactions tend to be reversible. For example, a reduced form of the nitroprusside ion can be protonated to give a blue solid:

$$[Fe(CN)_5NO]^{3-} + H^+ \rightarrow [Fe(CN)_5(NOH)]^{2-} \qquad (2)$$

However, in the $NO^-$-type complexes there is strong $\sigma$ donation and weak $\pi$ acceptance, so the nitrogen atom does not become electron-rich and is not susceptible to protonation.

There is another way in which presence of a nitrosyl group in a complex may also affect the other ligands in the complex. If another ligand is *trans* to the nitrosyl ligand, the two ligands will share bonding orbitals on the metal. When the nitrosyl is bound as $NO^-$, there is very strong donation of electron density *via* the $\sigma$ bond from the nitrogen to the metal. However, since the overlap of the d and π orbitals is very restricted, the metal cannot compensate for this extra electron density via π back-donation. Any ligand *trans* to this NO⁻ ligand would have to form a bond with the metal by donating even more electron density to the same σ-type orbital as the NO. Such a bond would be difficult to form. Indeed, many of the characterized complexes containing a formal NO⁻ ligand have a $d^6$ configuration (such as $Ru^{2+}$). They would be expected to be six-co-ordinate, octahedral structures, but most are only five co-ordinate and are square pyramidal in structure. The NO is always at the top of the pyramid, so there is no ligand *trans* to it. It is assumed that this is so because any ligand *trans* to the NO cannot be held to the metal. In contrast, five-co-ordinated complexes containing linearly bonded NO are generally trigonal bipyramidal, with the NO in one of the axial positions (Figure 9.3).

The description given above may help to understand how NO works in the body. When NO binds to iron porphyrins, the bond to the ligand *trans* to the NO is considerably weakened, and this ligand often dissociates altogether. Teddy Traylor at the University of California suggested that this is how NO activates the enzyme guanylate cyclase (sGC) (Chapter 1). The incoming NO molecule causes the haem complex to dissociate from the imadizole ligand at position 105 on the enzyme (Figure 9.4). This theory was put to an elegant test by Judith Burstyn (University of Wisconsin) and Thomas Spiro (Princeton University). They theorized that substitution of the Fe(II) atom by Mn(II) should render the enzyme inert to NO, because with one electron fewer at the metal centre there would be less *trans*-labilization produced when NO bound to the metal. On the other hand, when the Fe(II) was replaced by Co(II), the addition of an extra electron to the metal centre should make the *trans*-labilizing effect of the NO even more pronounced. The metal-substituted enzymes were produced and the experiment conducted. Sure enough, the sGC(Mn) showed no increase in activity upon addition of NO, whilst the sGC(Co) showed a higher level of activity.

**Figure 9.3**   *Complexes containing a bent NO ligand are usually five co-ordinate with the NO in the apical position, whereas five-co-ordinate complexes containing linear NO are usually trigonal bipyramical with the NO ligand in an axial position. Refer to the text for further explanation*

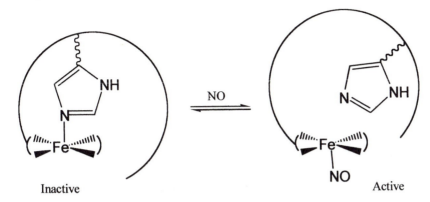

**Figure 9.4** *Proposed mechanism of activation of sGC by NO*

Although we cannot, at the moment, see all the implications, much of what is done by NO in the bodies of living things is a consequence of its behaviour as a ligand for transition metals. Few inorganic chemists would have thought that reactions carried out in laboratories are, in fact, also occurring in our bodies.

## FURTHER READING

G. Wilkinson, R.D. Gillard and J.A. McCleverty (eds), *Comprehensive Coordination Chemistry*, vol. 2, Pergamon Press, Oxford, 1987, 101.

J.A. McCleverty, Reactions of nitric oxide coordinated to transition metals. *Chemi. Rev.*, 1979, **79**, 53.

R. Eisenberg and C.D. Meyer, The coordination chemistry of nitric oxide. *Acc. Chem. Res.*, 1975, **8**, 26.

T.W. Hayton, P. Legzdins and W.B. Sharp, Coordination and organometallic chemistry of metal–NO complexes, *Chem. Rev.*, 2002, **102**, 935.

P.C. Ford and I.M. Lorkovic, Mechanistic aspects of the reactions of nitric oxide with transition-metal complexes, *Chem. Rev.*, 2002, **102**, 993.

*Chapter 10*

# Removing the Villain – Catalytic Converters

It came as a surprise to the residents of Los Angeles to discover that it was the motor car, not the local butadiene plant, that was responsible for the smog they were forced to endure. At the end of the Second World War disgruntled citizens established the Bureau of Smoke Control, paving the way for legalization to reduce exhaust emissions. In 1960 the Motor Vehicle Pollution Control Board was established in California. Its first act was to insist that cars be fitted with the first form of vehicle emission control technology in the USA, positive crankcase ventilation. This sounds impressive but all it means was that gases from the crankcase (the bottom part of the engine containing the pistons and crankshaft) were withdrawn and re-circulated through the combustion chamber.

This was just the beginning. In 1966 California adopted the first exhaust emissions standards in the world, limiting the amount of unburned hydrocarbons and carbon monoxide (CO) a vehicle could emit. The standards were made more stringent in 1970, and were joined by EU regulations on the other side of the Atlantic. At the time the laws were passed, there was no technology available that could meet the standards, and the motor industry protested. However, in a show of how much can be achieved when there is no choice, the motor industry responded by developing the catalytic converter.

The first catalytic converters were designed to deal with only CO and unburned hydrocarbons, hence they are termed two-way converters. The easiest way to eliminate these pollutants is to oxidize them:

$$2CO + O_2 \rightarrow 2CO_2 \tag{1}$$

$$\text{hydrocarbons} + xO_2 \rightarrow CO_2 + H_2O \tag{2}$$

It is a stroke of luck that both CO and hydrocarbons are thermodynamically unstable. The free energy of $CO_2$ is lower than that of CO; likewise the free energy of water and $CO_2$ is lower than that of hydrocarbons. The oxidation reactions do occur spontaneously, but at an incredibly slow rate because the energies of intermediate stages in the reaction are very high. A way of overcoming this large energy barrier is needed, and a catalyst system is ideal. The idea of such a catalyst goes back a long way, to a patent filed by G.P. Cross, W.P. Biller, D.F. Greene and K.K. Kearby of Esso (now ExxonMobil) in the 1960s. The rates of the reactions can be increased by many orders of magnitude if they are carried out on the surface of a metal, which allows the reaction to proceed *via* lower energy intermediates (Figure 10.1).

Many metals will catalyse these reactions, but the conditions in a motor car exhaust are quite special. The gases flow out of the engine at a predetermined (very fast) rate and temperature. Even with the maximum surface area that can possibly be contained in a little metal canister suspended underneath the front passenger seat, the outflow gases will spend little time on the surface of the catalyst. A very active catalyst indeed is required. Such a metal does exist, but unfortunately

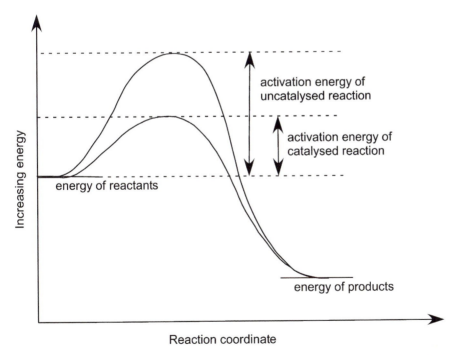

**Figure 10.1** *A catalyst increases the speed of a reaction by allowing it to proceed via lower energy intermediate states*

it is prohibitively expensive. The highest percentage of CO and hydro-carbons was oxidized using a mixture of platinum and palladium as the metal surface. These metals had the added advantage of not being deactivated by residual amounts of sulphur in the fuel. However, the lead antiknock compounds used in the fuel at that time were another matter. They had a disastrous effect on the metal surface and this is one reason why legislation to eliminate lead compounds from petrol was enacted. The catalyst metals were finely dispersed on a solid labyrinth of γ-alumina, which contains many channels for the exhaust gases to pass through. An excess of oxygen was ensured by placement of an air intake between the engine and the catalyst (Figure 10.2).

Two-way converters were first introduced in Los Angeles in 1975, and they did help to clean up vehicle emissions. There was just one problem: NO. Two-way converters did nothing to reduce smog forma-tion; if you oxidize NO you simply get $NO_2$ and thence smog and acid rain. Once again, the lawmakers were ahead of technology. NOx emis-sions were limited by legislation, with the motor industry given a few years to find the means. The best way to deal with NO is reduce it, giving $N_2$ plus another oxidized species.

It is again fortunate that NO is thermodynamically unstable. The enormous challenge was to find a way of reducing NO at the same time as oxidizing CO and hydrocarbons. Catalysts in sequence were tried, *i.e.* a reduction catalyst followed by an air intake and an oxidation catalyst. This was expensive, requiring two catalysts instead of one. In any case, under fuel-rich conditions ammonia was formed in the reducing catalyst. This was oxidized back to NO again in the oxidizing catalyst.

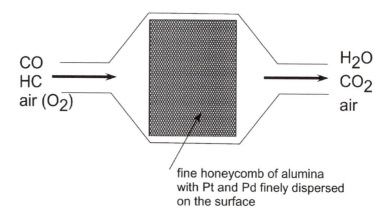

fine honeycomb of alumina
with Pt and Pd finely dispersed
on the surface

**Figure 10.2**  *Schematic diagram of a two-way converter*

When the source of the pollution is large and stationary, such as a power plant, there is a solution. A carefully controlled amount of reductant, such as urea or ammonia, can be added to the effluent to reduce the NOx, but it is not exactly practical to transport urea or ammonia around in a car.

Perhaps with the benefit of hindsight the solution to the problem is obvious, as there are two potential reducing agents being carried around in a car anyway, petrol and CO. If the car engine can be operated in such a way that all the oxygen in the intake is used up in burning all the fuel, the CO and NO can be removed by one catalyst. As it is oxidized, the CO reduces the NO. This kind of catalyst is called a three-way converter (Figures 10.3 and 10.4). The reaction is:

$$2NO + 2CO \rightarrow N_2 + 2CO_2 \qquad (3)$$

The balance is an incredibly fine one. If there is too much air, NO is oxidized. If there is too much fuel, the vehicle risks belching out CO, hydrocarbons and ammonia (Figure 10.5). To get the balance right, advances were necessary not only in chemistry but also in electronics. Electronic control processors and gas sensors were developed to give closed-loop engine control. A sensor placed in the engine outflow measures whether the exhaust gas mixture is net oxidizing (lean – too little fuel for the oxygen) or net reducing (rich – not enough oxygen to burn the fuel). This information is passed to a control unit, which in turn adjusts the amount of fuel sent to the fuel injectors. The control algorithms are now so sophisticated they can anticipate what will happen during uneven driving modes such as sudden acceleration.

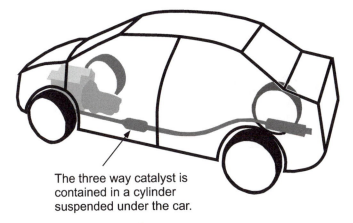

The three way catalyst is contained in a cylinder suspended under the car.

**Figure 10.3** *Position of the three-way converter in a car*

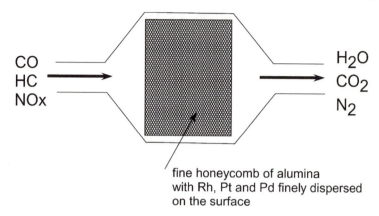

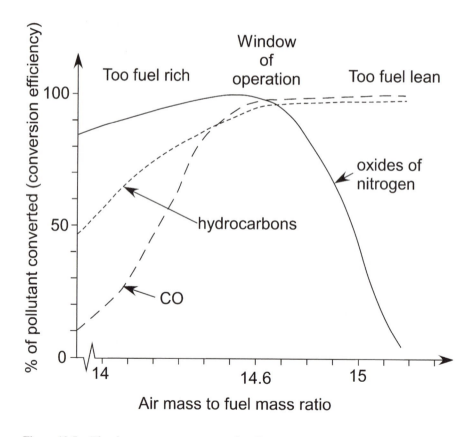

**Figure 10.4**   *Schematic diagram of a three-way converter*

**Figure 10.5**   *The three-way converter is only effective over a narrow air-to-fuel ratio*

However, no sensor can anticipate everything, and this is where the chemistry comes in.

Using an air:fuel ratio of about 14.65:1, formation of ammonia is avoided, firstly because the amount of available hydrogen present in the system is very small and, secondly, by carefully choosing a catalyst that is not very selective towards ammonia formation. The platinum/palladium catalysts used in the two-way converters did not catalyse NO reduction, but addition of rhodium to the mixture proved effective. The idea of adding rhodium to the catalyst formula was the breakthrough that allowed three-way converters to be developed. The credit goes to C. Meguerian, E. Hirschberg and F. Rakovsky of the then Amoco Corporation (now part of BP) for a patent filed in the 1970s. Rhodium is a by-product of platinum recovery and is thus rather scarce. Its introduction into three-way converters caused its price to skyrocket. There is nothing like expense to stimulate the search for an alternative, and three-way catalysts using only platinum/palladium have been reported. In these catalysts the amount of palladium used is far higher than in the two-way catalysts. Needless to say, exact catalyst formulations are closely guarded industrial secrets; many companies have several formulations available and can vary them according to the prevailing market prices of the metals.

The catalyst metals are very finely dispersed on a honeycomb of high surface area $\gamma$-alumina together with ceria ($CeO_2$) and zirconia ($ZrO_2$). Ceria would seem to be an exotic addition to the system, but it greatly improves the catalyst performance. Cerium switches very easily between its $Ce^{3+}$ and $Ce^{4+}$ oxidation states. The ceria acts as a sink for oxygen, absorbing it when oxygen is in excess and releasing it again when oxygen is scarce. The tiny particles of $CeO_2$ have a tendency to amalgamate, or sinter, over time in the heat of the system; the $ZrO_2$ is added to help prevent this.

Even though catalysts for three-way converters without rhodium do exist, they tend to have reduced activity for NO reduction. Why would rhodium, and only rhodium, be so active at promoting reduction of NO to $N_2$? Catalytic systems are notoriously difficult to study. They are put to industrial use because they are known to work, and it may be many years before someone finds out the precise reaction mechanism. As yet there is no probe that can be used to look inside a three-way converter and see exactly what the molecules inside do as the car drives around. The only recourse is to construct a model, which will be different in some way to the real system. Results from model systems give an insight into what may be happening in the real system.

One way to examine the mechanism is to look at what happens when NO interacts with a single crystal face of rhodium, platinum or palladium metal. Theoretical studies have been carried out using complex calculations to predict the shapes and energy levels of bonding orbitals when this happens. The molecular orbitals in a metal are extensively delocalized and extend across all the atoms present. This gives large numbers of orbitals with very similar energies, almost forming a continuum. The bonding orbitals will be filled to a greater or lesser extent depending of the number of d electrons the metal has. The highest occupied energy band is called the Fermi level, and it varies from metal to metal (Figure 10.6).

The Fermi level in platinum and palladium is *below* the energy of the lowest antibonding orbital of NO. This suggests that when NO attaches to a platinum or palladium surface there is a net flow of electron density out of the NO $2\pi^*$ orbital and into the metal. This in turn strengthens the NO bond. The converse is true for rhodium. The Fermi level of rhodium is *above* the $2\pi^*$ orbital of NO, so net electron density is expected to flow from the metal to the NO, thereby weakening the NO bond. NO adsorbed to rhodium is therefore more likely to dissociate than NO adsorbed to platinum or palladium. This is borne out by experimental evidence from infra-red spectroscopy and electron energy loss studies. NO is adsorbed on rhodium without dissociating, but only at low temperatures. At temperatures above 250K adsorption is dissociative. No dissociation occurs on platinum or palladium surfaces. If $N_2$ is to be produced from NO, the N–O bond must break at some point in the process. One theory argues that since rhodium dissociates NO, the mechanism must involve this as a first step, followed by pairing of N atoms on the surface of the rhodium and scavenging of an oxygen atom by CO. Unfortunately, high concentrations of NO interacting with a single crystal surface of a metal are a far cry from the highly dispersed metal particles on a support and the mixture of gases in a real three-way converter.

The reaction of CO with $O_2$ rather than NO has been studied both on a rhodium metal surface and on finely dispersed rhodium on a support. The size of the rhodium particles made no difference to the reaction, and results from the studies on the bulk rhodium could be used to predict what would happen on the dispersed rhodium. When the reaction of CO with NO using a rhodium catalyst was studied in the same manner, the results obtained with the dispersed rhodium could not be predicted from the results obtained with the bulk rhodium. The only way to make sense of them was to assume that the dissociation of adsorbed NO became slower as the rhodium particle size decreased. The

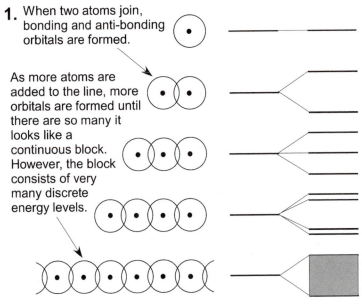

**1.** When two atoms join, bonding and anti-bonding orbitals are formed.

As more atoms are added to the line, more orbitals are formed until there are so many it looks like a continuous block. However, the block consists of very many discrete energy levels.

**2.** The s-orbitals in the atoms will overlap to form an s-band, whilst the p-orbitals form a p-band. If there is a difference in energy between these levels, it is called the band gap.

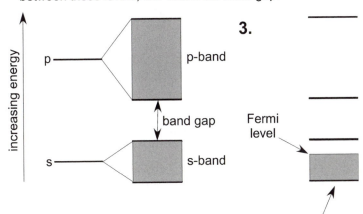

increasing energy

p — p-band

band gap        Fermi level

s — s-band

**3.**

Usually in a metal, these bonding orbitals are only partly filled. The energy level of the highest filled level is called the Fermi Level

**Figure 10.6**   *The Fermi level is the highest occupied energy band in a metal*

interaction between NO and finely dispersed rhodium on a support has been examined using infra-red spectroscopy. The most intense peak observed is at 1910 cm$^{-1}$, assigned to the Rh–NO$^+$ (nitrosonium) adsorbed state. In this state the NO has lost the electron from its $2\pi$ antibonding orbital and the NO bond strength is thereby increased, reducing the likelihood of N–O dissociation. This is not unlike the situation in which NO reacts with a number of transition metals in solution, as explained in Chapter 9.

Extrapolation between these kinds of infra-red studies and catalytic mechanisms is dogged with pitfalls, but such studies do give other evidence that is surely significant. When the interaction between finely dispersed platinum on a support and a mixture of NO and CO was studied, only the CO was observed to adsorb to the platinum. The platinum had such a high affinity for CO that the NO was crowded out. Perhaps platinum does not catalyse the reduction of NO with CO because in the presence of both gases, the NO never adsorbs to the platinum surface. When spectra of NO adsorbed on to finely dispersed, supported rhodium were compared with those of NO adsorbed on finely dispersed, supported palladium or platinum, the dinitrosyl species (ON–NO) was only observed on the rhodium.

Linking of two nitrogen atoms is crucial to the formation of N$_2$, so if a metal cannot facilitate this it will be a poor NO reduction catalyst. The infra-red evidence implies that in the three-way converter the mechanism of NO reduction is *via* the pairing of two NO molecules on the catalyst surface, followed by sequential loss of oxygen atoms and their uptake by CO, in contrast to the previous theory. Certainly N$_2$O has been shown to be an intermediate in the reduction of NO on finely dispersed, supported rhodium. The theoretical studies also lend support to this idea. The metal must release some electron density to facilitate the pairing of two NO molecules on its surface, and rhodium is in a better position to do this than platinum or palladium.

Time and effort will probably elucidate the mechanism of three-way converter action. In the meantime, the three-way converter is being developed further. Diesel engines pose another challenge in terms of emissions control. The three-way converter employed in petrol engines depend upon the engine operating with an air:fuel ratio of 14.65:1 or thereabouts. Diesel engines operate in a fuel-lean mode, leaving too much oxygen in the exhaust stream to allow reduction of NOx. Also, combustion conditions in diesel engines give rise to particulates. Particulates are a mixture of soot and liquid hydrocarbons left over from incomplete combustion of the fuel. The smaller particles, those less than 2.5 μm in diameter, have been directly linked to an increased risk of premature death. Strategies intended to reduce the amount of particulates produced, such as increasing combustion temperature, tend to increase the amount of NOx produced and *vice versa*. Currently, the best way to solve the problem is to find an exhaust treatment that reduces both the particulates and the NOx.

Particulates can be absorbed on to fine filters and thence oxidized to $CO_2$ and water. Typical exhaust outflow temperatures are 300–400 °C, too low for the oxygen in the exhaust gases spontaneously to oxidize the particulates. However, $NO_2$ is a stronger oxidant than $O_2$. A promising technology uses an oxidizing catalyst to convert NO into $NO_2$ before the filter. The $NO_2$ then oxidizes the particulates at the temperature of the exhaust outflow, regenerating NOx. The NOx might then be dealt with by selective catalytic reduction (described above), but there are numerous practical difficulties. Another means is *via* a so-called NOx adsorber. The NOx is passed over barium carbonate, and reacts with it to form barium nitrate:

$$BaCO_3 + NO_2 \rightarrow BaNO_3 + CO_2 \qquad (4)$$

When all the barium carbonate has been used, about every six seconds, a jet of fuel is added. The hydrocarbon temporarily creates a reducing environment over the adsorber, and the $NO_2$ is liberated and subsequently passed over a rhodium catalyst where it is reduced to $N_2$ (Figure 10.7). There are two major drawbacks to this procedure. Firstly there may be more particulates than $NO_2$ and, secondly, the oxidizing catalyst is poisoned by sulphur. Also, the oxides of sulphur react with barium carbonate to give barium sulphate. The barium sulphate must be purged periodically *via* a very high temperature engine cycle, with ramifications to the wear and tear on the engine. This technology can be used only with very low sulphur fuel.

The motor industry has another challenge besides diesel engines, so-called zero emissions. The main failing of the three-way converter is in its first 30 s or so of operation, while the catalyst is warming up to operating temperature. The emissions from this short time almost seem insignificant, until multiplied by the number of cars travelling to and from work, school or wherever each day. Low-temperature catalysts are being studied, as are ways of heating the catalytic converter more quickly. Other issues include finding catalysts to deal with the emissions from alternative fuels such as methanol. The big seven motor manufacturers will not be idle.

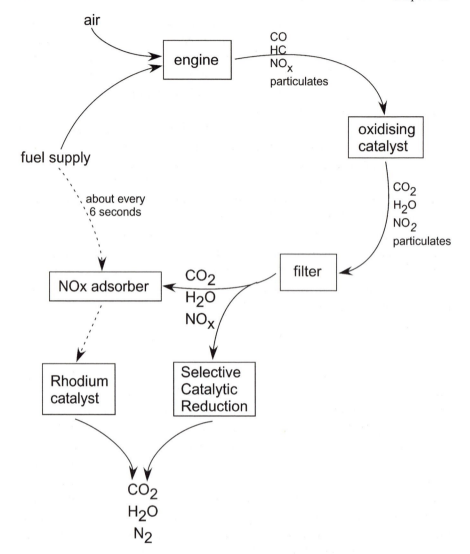

**Figure 10.7**  *A flowchart showing a promising technology for dealing with emissions from diesel engines*

## FURTHER READING

M. Shelef and R.W. McCabe, Twenty-five years after the introduction of automotive catalysts: what next? *Cataly. Today*, 2000, **62**, 35.

M. Shelef and G.W. Graham, Why rhodium in automotive three-way catalysts? *Catal. Rev.-Sci. Eng.*, 1994, **36**, 433.

H. Muraki and G. Zhang, Design of advanced automotive exhaust catalysts. *Cataly. Today*, 2000, **63,** 337.

W.A. Brown and D.A. King, NO chemisorption and reactions on metal surfaces: a new perspective. *J. Phys. Chem. B*, 2000, **104**, 2578.

R. Hoffmann, *The Same and Not the Same*, Princeton University Press, New York, 1995.

*Chapter 11*

# NO in Our Defences

It is time now to turn from the harm that NO in the atmosphere can do to the good that results from a measure of NO in human tissue. Our bodies are constantly being bombarded by foreign material. Air, dirt and water contain all sorts of life-threatening microbes: viruses, bacteria, parasites, fungi and non-self proteins. If any of these get into the tissues of the body the result may be dangerous and this has led to the development in all living things of very effective defence mechanisms. In mammals the skin and linings of the gut and lungs are very efficient at keeping out microbes. The skin is strong, flexible and constantly being renewed from below. Its glands produce an oily substance that can kill bacteria, as well as an enzyme, lysozyme, that can destroy bacterial cell walls. More is said about the skin in Chapter 14. In the respiratory tract some cells produce protective mucus, and others have hair-like projections, called cilia. The cilia beat in such a way as to propel mucus, and anything trapped in the mucus, up and through the throat and nose. The gastrointestinal tract also produces a protective mucus, while cells in the stomach release quantities of hydrochloric acid which, while produced largely to aid digestion, does also kill some harmful microbes. Unfortunately, it also irritates stomach ulcers.

The skin is covered with microbes, some of which are harmful, so it takes only a little cut or abrasion to compromise the protective role of the skin. The body is prepared for this and has complex internal defences constituting the immune system. This system is very sophisticated and designed to prevent the growth and proliferation of microbes. It consists of two parts: the natural or nonspecific system and the adaptive system. The adaptive system produces a response more or less selective for a particular microbe. For example, when smallpox viruses infect a person, the adaptive immune system produces an antibody with which to combat that particular virus. After the attack, the antibodies remained in the blood and so the victim has immunity from smallpox. This system is extremely complex and any further description is out of place here. The nonspecific immune system appears to be simpler, but in immunology nothing is simple. It is nonspecific in that it uses a number of strategies in an attempt to deal with *all* non-self matter, be it virus, bacterium, parasite or fungus. It is rather like an army with a variety of weapons using all of them to repel any invaders.

Several substances that have other roles in the body incidentally protect against invading microbes. The most important of these is a mixture of proteins called complement. If the invading bacteria do not have protective coats, complement can destroy them, but many harmful bacteria, because of their protective coats, evade this defence. A further defensive mechanism is a group of mobile scavenger cells. These cells can ingest, kill and digest foreign microbes. The process of ingestion is known as *phagocytosis*. The first group of such cells is known as the granulocytes because they have granules in them. Some granules contain powerful antimicrobial agents and others digestive enzymes. The second group of scavenger cells is called the macrophages and includes white blood cells. Both granulocytes and macrophages are produced by stem cells in bone marrow, and both act as scavengers by similar mechanisms, phagocytosis, killing and digestion, but there are subtle differences. Macrophages live longer, are larger and move around the body more slowly. Also, with macrophages digestion of the microbe is incomplete and this means that macrophages play a small role in the adaptive immune system. However, the process of killing microbes is common to both groups of cells.

Granulocytes and macrophages produce a collection of antimicrobial chemicals surprisingly like the household chemicals used for the purpose of hygiene. There is an enzyme in scavenger cells that produces hypochlorite ($ClO^-$), the principal component of domestic bleach. Oxygen in both granulocytes and macrophages is converted into a number of antimicrobial chemicals, the first of which is the superoxide radical anion ($O_2^{\cdot-}$). Other enzymes convert superoxide into hydrogen peroxide ($H_2O_2$), hydroxyl radicals ($HO^\cdot$) and singlet oxygen ($^1O_2$), and all are highly effective killers of microbes. It is an impressive array of toxic chemicals but each has only a brief life in the body. If that were not the case they would also kill the cells of the host. To ensure a brief life there are families of enzymes that convert the toxic chemicals into harmless products. For example, the enzyme superoxide dismutase (given the unfortunate acronym SOD) converts superoxide into hydrogen peroxide and this, in turn, is converted into oxygen and water by the enzyme catalase.

When dormant, granulocytes and macrophages do not produce toxic substances; activity comes only after stimulation. If there is a localized infection in the body both sets of cells migrate to the site and are stimulated to undertake phagocytosis and destruction of the microbes responsible for the infection. One method of stimulation is by means of a group of small proteins, called cytokines, which cells use to influence other cells, a process called cell-to-cell signalling. Cytokines, as well as other naturally occurring substances, stimulate macrophages to release toxic chemicals and thus kill invading microbes. One would have thought that the above list of these chemicals ought to be sufficient to kill all microbes, but there is at least one more and the story now takes a curious twist.

It had been known for many years that cattle with gastric infections excrete a lot of nitrate in their urine, but the reason for this was unknown. In the 1980s Steven Tannenbaum and his colleagues at MIT in Boston studied nitrate production in mammals in a series of carefully controlled laboratory experiments. They found that both mice and humans, maintained on a low and measured nitrate diet, excrete more nitrate than they eat. The origin of the extra nitrate and the reason for its production was unclear. Quite by chance they observed that urinary nitrate levels in one member of the human group in the study went up when he contracted fever and diarrhoea, just as had been observed in cattle.

Two American scientists at MIT, Dennis Stuehr and Michael Marletta, continued this work using cultured macrophages. They found that stimulation of cultured mouse macrophages by cytokines resulted in the release of both nitrate and nitrite in the culture medium. In 1985 they reported that bacterial infection in mice resulted in high nitrate levels in the urine and in 1987 they confirmed observations by John Hibbs and his colleagues in the University of Utah Medical Center (see Chapter 12) that nitrate and nitrite come from arginine. In the same year came the report from Moncada and his group, and Ignarro and his group of the arginine-to-NO pathway in the vasculature. Marletta immediately proposed that the same metabolic pathway for the production of NO exists in the immune system, and the presence there of an enzyme for production of NO was confirmed.

NO must be, therefore, another toxic chemical produced by macrophages as part of their defensive action. It was thought at the time that formation of nitrite and nitrate from NO was easy to understand as NO is readily oxidized to $NO_2$ and this reacts with water to give the two anions:

$$2NO_2 + H_2O \rightarrow HNO_2 + HNO_3 \qquad (1)$$

Subsequent work has shown that this may not be the exact pathway (see later) but there is now no doubt that the arginine-to-NO pathway exists in the immune system. The enzyme responsible for the production of NO in the immune system, iNOS (see Chapter 3), has been located, isolated and characterized.

There was great surprise in scientific circles when a role for NO was found in the vasculature. That the same process occurs elsewhere in living things seemed, at the time, almost incredible. The role of NO in smooth muscle relaxation is that of a benign messenger; in the immune system it acts as an aggressive, toxic agent. For NO to do both is a remarkable tribute to the unique chemistry of this, apparently, unremarkable molecule.

How does NO act as a toxic agent? The first answer given was that because it is a radical, it is very reactive and therefore destructive of cellular structures. It is now generally agreed that, although a radical, NO is not reactive enough to be particularly destructive. Much more NO is produced by macrophages and related cells than by endothelial cells but that alone still does not explain its toxicity. The answer to the question in any particular instance is not certain but there are a number of likely reactions of NO in living cells that could result in cell death. Before listing them we must distinguish between two types of cell death: *necrosis* and *apoptosis*. Necrosis is a sledgehammer approach. Destroy a crucial structure in a cell or interfere with a vital chemical process and the cell will die in a messy manner, leaving much unpleasant debris behind. Apoptotic cell death is a much more subtle process and one in which the cell commits suicide in a planned and orderly manner. It is also called programmed cell death. NO is probably involved in both necrosis and apoptosis; the former is described here and the latter in Chapter 12.

We will now look at some of the chemical reactions in which NO could participate and which might bring about necrosis.

1. One of the characteristics of radicals shared by NO is a ready reaction with other radicals. As well as producing NO, the immune system generates superoxide ($O_2^-$) and these two can react to give peroxynitrite, a species little known even to chemists:

$$NO + O_2^- \rightarrow O{=}N{-}O{-}O^- \qquad (2)$$

Peroxynitrite has two properties relevant to a role in the immune system. Firstly, it is a powerful oxidant and, secondly, it

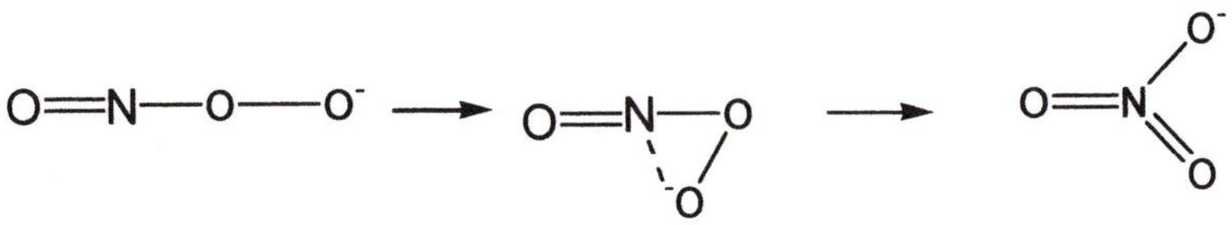

$$O{=}N{-}O{-}O^- \longrightarrow O{=}N{-}O \longrightarrow O{=}N{\overset{O^-}{\underset{O}{\bigvee}}}$$

**Scheme 11.1**   *Isomerization of peroxynitrite*

rapidly changes into nitrate, probably *via* a cyclic intermediate (Scheme 11.1). During its short life peroxynitrite may function as a toxic chemical by oxidizing, for example, the delicate cell membrane and thus killing the cell or by converting key thiols into disulphides. However, any excess of peroxynitrite soon becomes a benign species, nitrate, ideal for excretion in urine. Because of this rapid isomerization peroxynitrite is unlikely to harm the cells of the host. Peroxynitrite is also a radical nitrating agent and immune activity may result in the appearance of nitrated tyrosine (**11.1**) in some proteins. Nitration appears to be limited to tyrosine, since tyrosine readily forms a radical and nitration is a radical–radical reaction. Nitration of tyrosine within an enzyme deactivates the enzyme and could ultimately lead to cell death.

2. As described elsewhere (Chapter 8) oxidation of NO in solution leads to the formation of $N_2O_3$. This is a nitrosating species and can react with cellular thiols to give *S*-nitrosothiols:

$$RSH + N_2O_3 \rightarrow RSNO + HNO_2 \tag{3}$$

Some microbial enzymes contain a thiol group at the active site, the part of the enzyme where the action occurs. Thus the generation of $N_2O_3$ near an enzyme may result in the enzyme being deactivated and prevent essential processes occurring within the metabolism of the microbe. The result is death of the microbe.

Before continuing this list of reactions it is necessary to digress to describe an important structure within a cell: the mitochrondrion. This

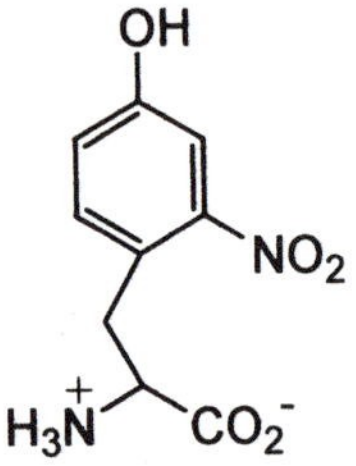

**11.1**

is an oval-shaped structure about 2 μm in length and 0.5 μm in diameter when mature (Figure 11.1). It has a complex internal structure, as shown by George Palade at The Rockefeller University and Fritjof Sjöstrand at Karolinska Institutet in Stockholm. There are two compartments: in the intermembrane space a process known as oxidative phosphorylation occurs while in the matrix reactions involved in the fatty acid and citric acid cycle take place. In oxidative phosphorylation oxygen is consumed, adenosine triphosphate (ATP) is formed and energy is released. Both oxidative phosphorylation and the citric acid cycle involve complex molecules containing iron–sulphur clusters. These curious structures consist of clusters of iron and sulphur atoms (principally 2Fe–2S and 4Fe–4S) as a stable unit. The sulphur may be inorganic sulphur or the sulphur of the thiol group of cysteine (Figure 11.2). Their role in oxidative phosphorylation is as part of the electron transfer chain. Two iron–sulphur cluster compounds have been described already (Chapter 6): Roussin's black salt and Roussin's red ester. As well as iron–sulphur clusters, other metal-containing compounds play essential roles in oxidative phosphorylation.

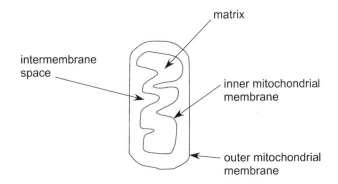

**Figure 11.1**   *Diagrammatic representation of a mitochrondrion*

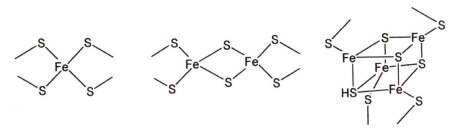

**Figure 11.2**   *Iron–sulphur clusters*

We can now return to our list of reactions that might explain the cellular toxicity of NO.

3. In a healthy cell NO plays a part in controlling the consumption of the oxygen (respiration) in mitochrondria but, under certain circumstances, binding of NO to the iron of an iron–sulphur cluster could cause disruption of the cluster. If this happens respiration ceases and the cell dies. As explained in Chapter 9 NO binds to iron very readily.

4. Oxidative phosphorylation also involves reactions at cytochrome c and $c_1$, which have iron-containing haem structures (Figure 11.3). NO could bind here, and the result would again be cell death.

5. The four-electron reduction of oxygen by cytochrome also involves copper ions. NO binds to copper, possibly not as strongly as to iron, but the formation of copper–nitrosyls would be disruptive.

6. An important enzyme in the citric acid cycle is aconitase (Figure 11.4), which contains a 4Fe–4S cluster. Binding of NO to any of these irons could result in cell death.

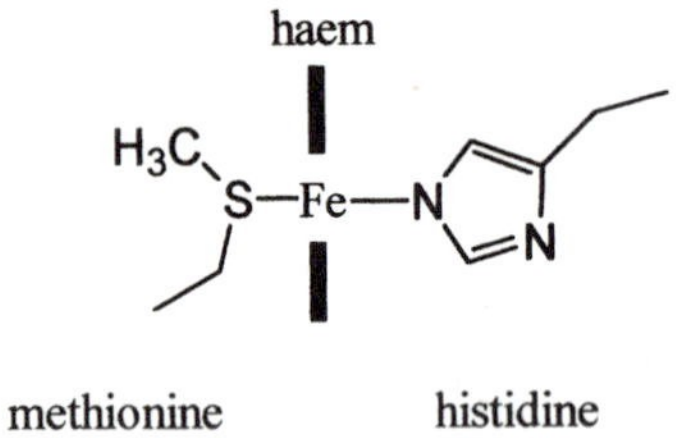

**Figure 11.3**  *Haem of cytochrome c has an iron atom bonded to histidine-18 and methionine-80*

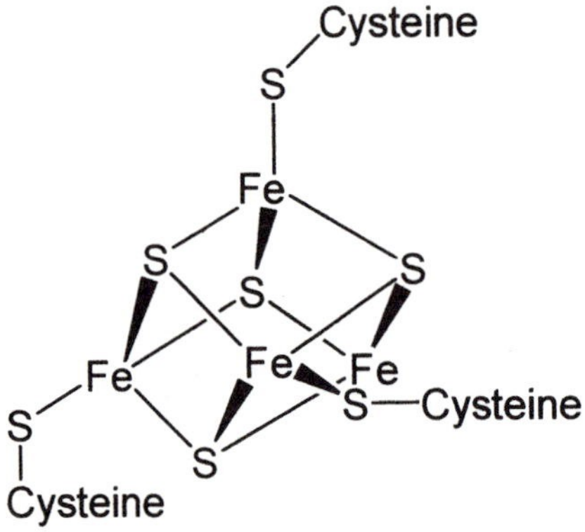

**Figure 11.4**  *The active site of aconitase has a 4Fe–4S cluster*

7. Some proteins contain zinc and sulphur in regions known as zinc fingers. Reaction of an oxide of nitrogen with sulphur results in destruction of the fingers and seriously affects the function of the protein.
8. NO can interact with several different groups on the enzymes (ribonucleotidases) that effect DNA synthesis, effectively inhibiting them.

It is doubtful if this list is exhaustive. The success of NO as a toxic agent is largely because so many living processes involve inorganic chemicals (*i.e.* metal ions). If biological chemistry involved only the chemistry of C, H, O, P and S then NO would be redundant. Our metabolic processes are acting out the information contained in Chapter 9.

The origin of the nitrite and nitrate observed by Stuehr and Marletta can now be discussed. The nitrate could come from the reaction of NO with superoxide. Also, any excess of $N_2O_3$ in solution, formed by oxidation of NO, reacts with water to give only nitrite:

$$N_2O_3 + H_2O \rightarrow 2HNO_2 \qquad (4)$$

Nitrite is also quite readily oxidized to nitrate and so finding both nitrite and nitrate in the culture medium of stimulated macrophages is not surprising.

It shows commendable economy on the part of Nature to chose the same molecule (NO) to do two jobs in the body but does this ever lead to difficulties? There is a medical condition known as septic shock characterized by very general and widespread infection. The victim often dies, not from the infection but from lowered blood pressure. This could be caused by excessive amounts of NO produced by macrophages, resulting in very general rather than localized dilation of blood vessels, with consequent lowering of blood pressure. We are more familiar with the hazards of high blood pressure but blood pressure that is too low is equally dangerous. There are reports of patients being snatched from the jaws of death by administration of inhibitors of the enzyme that produces NO. Blood pressure is thus brought under control and the infection is cured with antibiotics. Unfortunately, this treatment is not always successful, for reasons that are not well understood, and septic shock remains a major cause of death, particularly in the developing world.

The above account of the role of NO in the immune system is nothing like the full story and much remains to be elucidated. Quite large quantities of NO are produced by macrophages and it might seem surprising that its presence remained undetected until the 1980s in spite of so much scrutiny of human physiology by scientists. One reason for this is the difficulties experienced in detecting NO and another is the ease with which it is converted into nitrate, a species that could equally well come from the metabolism of protein. Although NO is small, and has a short life, its impact on infection and inflammation is enormous.

There is one infection where NO has a rather special role. Malaria is, once again, a major world health problem, particularly for children. In Africa alone over a million children a year die of malaria. Worldwide the total number of deaths from malaria exceeds two million and many more suffer from the debilitation associated with the disease. At one time it was thought that the disease could be eradicated, in the way that smallpox has been, but the strategies implemented by the World Health Organization to this end in the 1950s proved unsuccessful. So unsuccessful, in fact, that there is now more malaria than ever.

Malaria is caused by a tiny parasite (*Plasmodium*) entering the blood-stream when an infected female mosquito bites the victim in order to suck blood. The life cycle of *Plasmodium* in humans is very complex, involving both sexual and asexual stages of reproduction in both the blood and the liver. Within the bloodstream *Plasmodium* enters a red blood cell, reproduces asexually, the cell bursts and many young para-sites are released into the bloodstream to infect even more red blood cells. The intermittent nature of the fever, so characteristic of malaria, coincides with the process of reproduction. The destruction of red blood cells is the cause of anaemia and lethargy in the victim. There is also a toxin produced by the parasite, which is harmful. Red blood cells containing the parasite have a tendency to adhere to the endothelial cells lining the blood vessels in much the same way that platelets adhere to vessel walls (see Chapter 2). In the most serious form of malaria, cerebral malaria due to infection by *Plasmodium falciparum*, this adhesion occurs in the arteries of the brain and may lead to a complete blockage of the artery and death of the victim. Cerebral malaria is frequently fatal.

For many years quinine, obtained from the bark of the cinchona tree growing naturally in Peru, and its synthetic cousin chloroquinine, were used for the treatment of malaria. The widespread use of quinine in the nineteenth century allowed Europeans to explore tropical countries and the spread of the British Empire owes much to this humble drug. Equally the triumph of the Allies in the Pacific in the Second World War owes much to the use of chloroquinine as at one point more soldiers were being withdrawn from the frontline because of malaria than because of battle casualties. One reason for the recent increase in deaths from malaria is that *Plasmodium*, particularly *P. falciparum*, has become resistant to chloroquinine and there is now no really effective and readily available drug for the treatment of cerebral malaria. Because of this crisis, there is so much talk of antimalarial drugs that the body's natural defence against malaria is easily overlooked. This defence involves, needless to say, the production of NO, but its role is rather different from that in a bacterial infection.

The adaptive or specific response of the immune system to malarial infection is short-lived. It is also complex because there has to be a different set of antibodies for each stage in the life cycle of *Plasmodium*. The response of the nonspecific immune system to *Plasmodium* is much better understood. Malarial infection induces the release of cytokines, which result in the formation of a number of toxic agents, including NO from iNOS. NO is quite effective in killing *Plasmodium* when it is in the bloodstream but not once it has invaded the red blood cells. The blood stage outside the red blood cells is a minor one in the life of *Plasmodium*. The death of the parasite at this stage is hardly likely to curb an infection and cytokines may also have a harmful effect. Overproduction of cytokines contributes markedly to the diseased state, causing fever and acidosis (an unpleasant condition in which the body produces two abnormal acids, leading to lassitude and vomiting). However, there is a very positive side to NO production during malaria. It inhibits one of the most serious aspects of the disease. In much the same way as NO inhibits platelet adhesion to endothelial cells (see Chapter 2) NO inhibits the adhesion of red blood cells that are infected with *Plasmodium* to blood vessel walls and blockage of vessels is therefore less likely. Death from cerebral malaria may arise when the victim has an inadequate NO response, despite high levels of circulating cytokines. There is some evidence that the severity of the disease correlates with the unusual nature of the gene producing iNOS. In Tanzania, where malaria is rife, death from the disease is most common in children over one year of age and below five. This parallels NO production, which declines after the first year of life but rises again after five. A drug therapy based on NO production may not kill the malarial parasite but may lessen its ability to cause death. Curiously, the mosquito can produce an iNOS enzyme that is very similar to the human one; whether it uses iNOS as part of its defence against *Plasmodium* is a matter of speculation.

## FURTHER READING

M.A. Marletta, P.S. Yoon, R. Iyengar, C.D. Leaf and J.S. Wishnok, Macrophage oxidation of L-arginine to nitrite and nitrate: nitric oxide is an intermediate. *Biochemistry*, 1988, **27**, 8706.

F.Y. Liew and F.E.G. Cox, Nonspecific defence mechanism: the role of nitric oxide. *Immunol. Today*, 1991, **12**, A17.

K.A. Rockett, M.M. Awburn, W.B. Cowden and I.A. Clark, Killing of *Plasmodium falciparum in vitro* by nitric oxide derivatives. *Infection and Immunity*, **1991**, 3280.

S.L. James, Role of nitric oxide in parasitic infections. *Microbiol. Rev.*, **1995**, 533.

*Chapter 12*

# NO and Tumour Growth

Control of cell growth and division is an important but little understood aspect of mammalian physiology. Organs grow to a predetermined size and then growth stops. In most organs there are special
reserve or stem cells that are capable of growing into organ-specific cells
in response to, say, injury. In organs consisting of very highly differen-
tiated cells, particularly brain cells, the capacity to grow new cells is
lost in adulthood and this explains why brain damage is so serious.
In contrast, there are tissues, such as the intestine, blood and the
immune system, in which there is rapid turnover of cells. The process
of cell replacement is controlled by a number of stimulatory and inhibitory factors. In healthy tissue a delicate balance between stimulation
and inhibition is established, but when repair is required the balance
changes and stimulation predominates. If there is a net increase in
the number of cells due to over-stimulation, the condition is known as
hyperplasia and this is normally reversible. If, however, hyperplasia gets
out of control and becomes irreversible then a tumour may result.

It is not easy to define exactly what a tumour is but, broadly,
tumour cells are characterized by not responding to the normal growth-
controlling mechanisms of the body. A benign tumour may arise in
any tissue and grow in such a way as to cause damage by pressure
or obstruction but such tumours do not spread to other sites. *In situ*
benign tumours usually develop in surface tissue and are generally
small. They are cancerous in that the cells of the tumour differ from the
cells from which they were derived but they do not invade any tissue
below the surface. Fully developed cancerous cells differ from the cells
from which they were derived, proliferate rapidly and invade the surrounding tissue and may eventually break off and travel in the blood
to distant sites (metastasis). What causes cells suddenly to behave in
this deviant way is not fully understood but some factors have been

identified. Carcinogenic (cancer-causing) chemicals in the environment, such as cigarette smoke, is one such factor and the role of sunlight in causing skin cancer is described in Chapter 14. Cancer is mainly, but not exclusively, a disease of age and as life expectancy has lengthened so much in recent years there has been a dramatic rise in the incidence of cancer. Treatment of cancer has improved out of all recognition, partly because of early detection through mass screening, but it is still a major cause of death, more so in men than in women. Cancer is rarely the direct cause of death unless it invades a vital organ such as the liver. Rather it can precipitate anaemia and wasting, leading to greatly reduced resistance to infection.

The body does not give in to cancerous growth without a fight and the cells of the immune system (see Chapter 11) can destroy small tumours. Indeed, a rapidly growing tumour contains many macrophages doing battle with the proliferating cancer cells. If the immune system cannot keep pace with the rate of increase in the number of cells then a tumour becomes established. Tumour cells can be cultured in the laboratory and in the early 1980s it was known from *in vitro* experiments that macrophages, when suitably stimulated, kill tumour cells. Before a biological role for NO had been thought of, John Hibbs and his colleagues at the University of Utah Medical Center showed that this toxic effect was due to a chemical reaction in the macrophages in which L-arginine is converted into L-citrulline, nitrite and nitrate. Neither L-citrulline nor nitrite/nitrate is particularly toxic and so they concluded that there was still something to be discovered about the process. Their words are worth quoting:

"The actual mechanism of metabolic inhibition by products of the L-arginine-dependent effector mechanism remains to be defined. However, nitrite and oxygenated intermediates in the pathway of nitrite and nitrate synthesis could participate....."

Their words were prophetic for, in the following year (1987), the first accounts of the L-arginine-to-NO pathway in the vasculature were published and immediately suggested to Hibbs that the toxic agent produced by macrophages is NO. The nitrite and nitrate detected in their experiments was the result of NO hydrolysis and oxidation. In 1988 they confirmed that NO is, indeed, the toxic agent and suggested that the targets for NO attack are the iron atoms involved in cell respiration, the enzyme aconitase (part of the citric acid cycle) and DNA synthesis. These suggestions parallel almost exactly those of Marletta on the antibacterial action of NO (see Chapter 11). How far NO is involved in the body's defence against cancerous cell growth has yet to be determined and we are still a very long way from harnessing the toxic effect of NO in an anti-cancer drug.

Further research has confirmed that the ways in which NO kill tumour cells are very similar to its toxic action towards pathogenic bacteria, detailed in Chapter 11. This type of cell toxicity, which is purely chemical in nature, is known as necrosis. It is messy and the dead cells, before they have been removed by phagocytosis, may cause inflammation. There is another type of cell death, known as apoptosis or programmed cell death, discovered by John Kerr and his colleagues at the University of Queensland in 1972. It is sometimes called cell death Swiss-style because it is so well organized. There is a gene in cells which, when stimulated, can effect suicide of the cell in a very orderly and clean manner. It was known to occur in embryonic development but the suggestion that it also occurs in the daily maintenance of a mature organ was new. The rather unusual word apoptosis comes from the Greek 'falling off', as leaves in autumn. The dying cell has a high concentration of calcium ions and there is breakdown in the organization of its DNA. As the cell deteriorates the components are wrapped in protective protein shells and this prevents the leakage of potentially harmful material. Eventually the apoptotic cells are removed through phagocytosis by macrophages and neighbouring cells. Once the gene or genes controlling apoptotic death have been stimulated the process proceeds spontaneously. If there is a mutation in this gene the cell that habours it may fail to respond to the cue to die and proliferate uncontrollably. In a most exciting development in the NO story, some experimental evidence has suggested that NO can stimulate the gene for apoptosis. If further work confirms this role for NO, and delineates the circumstances under which it may occur, then we have yet another important function for this apparently unassuming and elusive diatomic molecule.

Much has been said about the isoform of NOS most commonly associated with blood vessels, eNOS, but blood vessels may also produce iNOS. Blood vessels supplying tumours have higher than normal levels of iNOS, possibly as a response to the inadequate arginine levels in the rapidly growing cells of the tumour. Administration of a NOS inhibitor *in vivo* leads to a collapse in tumour blood supply and results in cell death.

So far it would appear that the activity of NO in relation to cancerous growth is beneficial but, as with so many aspects of the NO story, it may not be that simple. Under oxidizing conditions in solution NO forms $N_2O_3$, a species that can nitrosate at physiological pH. There are

$$\underset{R}{\overset{R}{>}}NH \quad + \quad N_2O_3 \quad \longrightarrow \quad \underset{R}{\overset{R}{>}}N-NO \quad + \quad HNO_2$$

**Scheme 12.1** *Formation of a secondary nitrosamine*

$$\underset{CH_3}{\overset{CH_3}{>}}N-NO \quad \xrightarrow{\text{enzyme}} \quad \underset{CH_3}{\overset{HOCH_2}{>}}N-NO \quad \longrightarrow \quad HCHO \quad + \quad CH_3NNOH$$

$$CH_3NNOH \quad + \quad H^+ \quad \rightleftharpoons \quad CH_3-N_2^+ \quad + \quad H_2O$$

$$\Big\downarrow DNA$$

methylated DNA

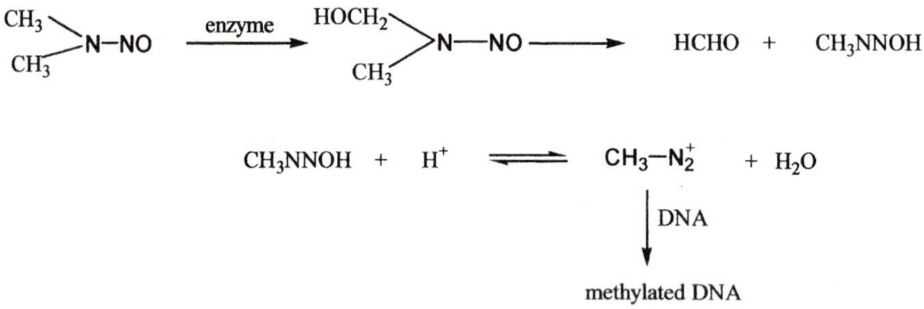

**Scheme 12.2** *Reaction between N-nitrosodimethylamine and DNA*

secondary amines present in tissue and they may be nitrosated to give secondary *N*-nitrosamines, known to be very powerful carcinogens (Scheme 12.1). They are carcinogens because they can chemically modify DNA and the way in which *N*-nitrosodimethylamine is thought to do this is shown in Scheme 12.2. How readily secondary nitrosamines form in living tissue is not known but it has been observed *in vitro* that activated macrophages will nitrosate morpholine to form *N*-nitrosomorpholine. Even if there just the possibility of *N*-nitrosamine formation the relationship between NO and cancerous growth is rendered ambivalent.

The possibility of amine nitrosation when sodium nitrite is used to prevent the spoilage of tinned meat by the bacterium responsible for botulism, *Clostridium botulinum*, is a matter of grave concern. Nitrite is very effective in inhibiting *C. botulinum* and has been used for this purpose over many years. Nitrite is not a nitrosating agent at physiological pH, but the stomach is much more acidic and the nitrite in tinned meat might well effect amine nitrosation there. There are experimental data from work with animals to show that urinary *N*-nitrosamine levels are elevated if the animal is fed on tinned meat. Whether this results in the increased incidence of cancer is not known. As far as humans are concerned there is no evidence to link eating tinned meat with the incidence of cancer and so nothing has been done to change the situation, in spite of the theoretical danger. A bonus in the use of nitrite as a preservative is that the attractive pink colour of tinned meat is due to the nitrosation of the myoglobin found in animal tissue.

*N*-nitrosamine formation is not the only danger from NO. Cancerous growth may be caused by chemical changes in the DNA and such mutations could result from the nitrosation of the heterocyclic bases in DNA. Some human cancers are associated with, among other things, the replacement of some cytosines on the DNA strand by thymines. The conversion of one base into another is a multistep process and one step could be nitrosation, as shown in Scheme 12.3. Much remains to be elucidated in this area but it may be that it is the amount of NO present in tissue that decides whether its effect is harmful or not. There is good experimental evidence to suggest that repeated infection of the liver, as in hepatitis, which causes the frequent activation of iNOS and the production of large quantities of NO, can lead to cancer of the liver.

As there is so much NO produced in or near cells it may seem surprising that DNA is not regularly and frequently affected by nitrosation. However, in a healthy cell any damage to the bases (the estimate is that there are many thousands of damaged bases produced in every cell during the course of a day) is immediately repaired by a very efficient and complex repair mechanism. It is only when the repair mechanism ceases to function properly that permanent mutations in DNA occur. Even then the probable outcome will be apoptosis of the cell but DNA mutation *can* lead to cancerous growth.

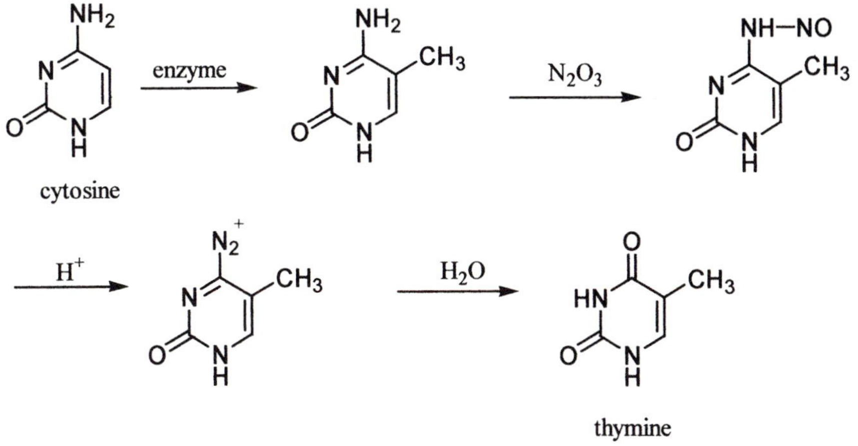

**Scheme 12.3**  *Conversion of cytosine into thymine by nitrosation*

The dual role of NO in the provenance of cancer is well illustrated by experimental data on breast cancer from Salvador Moncada and his group at University College London. Examination of samples of breast carcinomas showed a general elevation in the activity of both iNOS and eNOS. On the other hand, in experiments with cultured cancer cells it was found that certain breast cancer cell types were very susceptible to NO-mediated tumour killing. These experiments show that the concentration of NO plays an important role in the apparently paradoxical effects of NO on tumour cell biology. Clarification of the situation, no small challenge, could have enormous consequences for the treatment of certain types of cancer.

## FURTHER READING

J.B. Hibbs, R.R. Taintor, Z. Vavrin and E.M. Rachlin, Nitric oxide: a cytotoxic activated macrophage effector molecule. *Biochem. Biophys. Res. Commun.*, 1988, **157**, 87.

G.D. Kennovin, D.G. Hirst, M.R.L. Stratford and F.W. Flitney, Inducible nitric oxide synthase is expressed in tumour-associated vasculature: inhibition retards tumour growth in *The Biology of Nitric Oxide*, S. Moncada, M. Feelisch, M.R. Busse and E.A. Higgs (eds), Portland Press, London, 259.

T. Nguyen, D.V. Brunson, C.L. Crespi, B.W. Penman, J.S. Wishnok and S.R. Tannenbaum, DNA damage and mutation in human cells exposed to nitric oxide *in vitro*. *Proc. Natl. Acad. Sci. USA*, 1992, **89**, 3030.

D.C. Jenkins, I.G. Charles, L.L. Thomsen, D.W. Moss, L.S. Holmes, S.A. Baylis, P. Rhodes, K. Westmore, P.C. Emson and S. Moncada, Roles of nitric oxide in tumor growth. *Proc. Natl. Acad. Sci. USA*, 1995, **92**, 4392.

J.F.R. Kerr, A.H. Wyllie and A.R. Currie, Apoptosis: basic biological phenomenon with wide-ranging implications in tissue kinetics. *Brit. J. Cancer*, 1972, **26**, 239.

*Chapter 13*

# Bones, Joints and NO

Bone is a living tissue consisting of a protein structure upon which calcium salts, particularly calcium phosphate, have been deposited. Children manifest two periods of rapid bone growth, one during the first two years of life and the second during puberty. In contrast, brain growth is complete by the age of five. After rapid bone growth, in both length and diameter, during puberty the skeleton comes to a static state. However, bone renewal, as distinct from bone growth, occurs throughout life because bone cells have only a limited life and are continuously replaced by new ones. A wave of bone resorption (loss of bone cells) is followed by a period of deposition and the net result is complete turnover of bone, while preserving total bone mass. Two special cells are involved in these processes: the cells responsible for resorption are called osteoclasts and those involved in deposition are known as osteoblasts. Both are derived from bone marrow. Perturbation of the two processes is very serious. In rheumatoid arthritis and osteoporosis there is net bone loss because resorption exceeds deposition.

The control of bone resorption and deposition is a matter of finely tuned balance and with healthy bone the two processes are fully synchronized. By means yet unknown osteoclasts and osteoblasts do communicate and this is part of the synchronization mechanism. However, many external substances, such as 1,25-dihydro vitamin D3, cytokines (see Chapter 11) and oestrogen, play important parts in the generation, growth and activity of osteoclasts and/or osteoblasts. Because we had become used to such discoveries, it came as no great surprise when, in 1991, it was announced that NO was implicated in bone renewal in both healthy and diseased states. The situation is, at the time of writing, not fully understood and much more research is required to uncover the full picture. As is common in a topic where knowledge is advancing

103

rapidly, there are inconsistencies between reports in the scientific literature, partly because different researchers use different animal models.
Also, results obtained using cultured cells may or may not apply to
bone cells in living things. What follows is a summary of the current
position.

The process of resorption (bone depletion) is fascinating. Several
young osteoclasts merge to give multinucleated cells, these cells attach
themselves to bone and then each cell spreads out to cover a fairly large
area. The cell produces hydrochloric acid and a collection of enzymes;
this brew destroys bone immediately below the cell. Eventually, a pool
of liquid containing a high concentration of calcium ions derived from
bone forms. This high concentration outside causes an elevation of calcium ion levels inside the osteoclast by diffusion, resulting in activation
of the enzyme eNOS. The NO formed acts as a messenger molecule to
bring about detachment of the osteoclast from the bone surface, movement to a fresh site, reattachment and repetition of the process. Readers
may remember that eNOS in the vasculature is activated by the same
process, the elevation of calcium ion levels (see Chapter 1). Eventually
the osteoclast, after effecting bone loss at a number of sites, dies. Thus
NO plays a crucial role in the process of bone resorption. But too much
NO, as occurs when there is inflammation of a joint (see later), inhibits
replication of very young osteoclasts and also accelerates cell death of
older ones, destroying them before they have completed their task of
bone cell resorption. Resorption, if it is followed by bone deposition, is
an entirely beneficial process in maintaining healthy bones. The fact
that low concentrations of NO aid, and high concentrations hinder, the
process of desorption is another example of the fine balance required in
the production of NO for the healthy working of the body.

The effect of NO on osteoblasts, the cells responsible for bone deposition, is far less clear. If cultured osteoblasts are stimulated by
cytokines, iNOS is formed and quite large quantities of NO result, generated over a 48-hour period. The significance of this observation, and
whether the same effects occur in living things, are matters still to be
clarified. In addition to the production of NO by osteoblasts themselves, other cells in the bone environment produce NO and, although
the fine details have still to be elucidated, it is clear that this NO has an
effect on osteoblast development and activity. As with osteoclasts, what
is required by osteoblasts for success in bone deposition is production
of an optimum amount of NO.

Bone has an extensive vascular system and NO formed in the
blood vessels may have an effect on bone growth. Under normal
circumstances, with an upright posture, capillary blood pressure is at its

lowest in the bones of the head and maximal in the bone blood vessels of the feet. However, during space travel, with its weightlessness, this difference disappears. Studies have shown that space travellers lose bone mass in the feet and gain it above the heart. The explanation of this may be that the flow of blood in capillaries generates shear stress (see Chapter 1), a process known to result in release of NO from vascular endothelial cells. Weightlessness decreases shear stress generation of NO in the bone vessels of the leg and increases it in those of the head, thus stimulating osteoclast and osteoblast activity differently from that during terrestrial existence. The result is a changed pattern of bone growth.

The increase in blood flow, and associated shear stress generation of NO, occurring during exercise may explain the commonly noted increase of bone mass amongst athletes in serious training. Shear stress related NO production is probably not limited to the blood supply as there is also fluid flow in bone. Bone is quite a porous material. Whatever the details, it is clear that lack of exercise has serious consequences for bone renewal.

A particularly worrying aspect of bone renewal is its decrease in postmenopausal women due to a fall in oestrogen production. The reasons for this are little understood in spite of much research. It may be that oestrogen stimulates NO production from cells in the bone environment. If this is the case then the way is much clearer for an eventual treatment of osteoporosis. In the meantime, increased exercise may be the best treatment. The age to take up marathon running may be that of the menopause.

One of the many factors that discourages people from taking exercise is inflammation of the joints. There is a connection between infection and inflammation, and NO is part of that connection. Infection results in the release of a number of toxic substances by mobile cells of the immune system (see Chapter 11) but that may not be everything that happens. Infection, as well as physical and chemical damage, may cause changes to the tissues of the host, a process known as inflammation. The general characteristics of inflammation are heat, redness, swelling and pain, effects commonly seen with an infected wound or a minor blow. The heat and the redness are due to dilation of blood vessels in the area of damage or infection. The walls of the blood vessels change and become more permeable to proteins present in blood plasma, such that they leak out of the blood vessels and gather in the damaged or infected area. Proteins can become extensively hydrated and the build up of water in the damaged or infected tissue results in swelling and, possibly, pain. All these processes occur before arrival of the macrophages and related cells. When these do arrive they release the toxic chemicals detailed in Chapter 11 and foreign bacteria are killed, ingested and digested. Eventually the tissue is restored to its normal state. If there are no foreign microbes and only physical or chemical damage, macrophages fulfil the role of ingesting the debris of dead and dying cells, a necessary step in the rebuilding of normal tissue. Although the production of toxic chemicals by macrophages is, in general, beneficial, they may harm the host. Pain is enhanced because of the effect they have on nerve endings and they may also damage the tissues of the host. For example, whooping cough is caused by a bacterium *Bordetella pertissis* infecting the upper respiratory tract. There is now good evidence that the irritation to the tract responsible for the characteristic bouts of coughing is linked to the production of large amounts of NO, which damages the cells. *B. partussis* is very resistant and requires much activity on the part of the immune system before it is eradicated, and so the bouts of coughing continue for some weeks and cease only when the infection has been completely defeated. Although troublesome in some ways, inflammation may be seen as part of the healing process.

There is, however, a somewhat puzzling manifestation of inflammation: it can occur with no external stimulus. The immune system may respond to an internal factor as if it were foreign. The mobile scavenger cells of the immune system are mobilized, they congregate in one particular region and produce, in a series of programmed events, the toxic agents normally released to kill invading microbes. Under these circumstances all that can happen is damage to the tissues of the host, giving rise to chronic inflammation (inflammation over a long period). Rheumatoid arthritis (RA) is an example of such an autoimmune inflammatory disease affecting joints (Figure 13.1). It affects many people of all ages and is a major debilitating condition. What causes the immune system to behave in such an unprofitable way is not known and treatment of the symptoms is difficult and costly. The principal symptoms are joint pain and swelling, excessive amounts of synovial fluid (the fluid that lubricates the joint) and cartilage destruction (Figure 13.2). Macrophages and other mobile cells migrate to the joint and congregate in the synovial fluid, releasing large quantities of NO. There are elevated levels of nitrite in the synovial fluid, and possibly in the serum, of sufferers from RA. Nonspecific inhibitors of NOS can dramatically suppress development of joint inflammation in some animal models, and genetically engineered mice without the gene for producing iNOS (known as 'knockout' mice) show a much reduced tendency to develop joint inflammation. The evidence that NO is

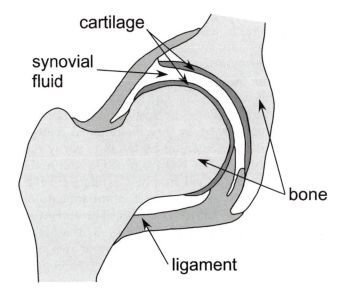

**Figure 13.1** *Simplified diagram of a healthy joint*

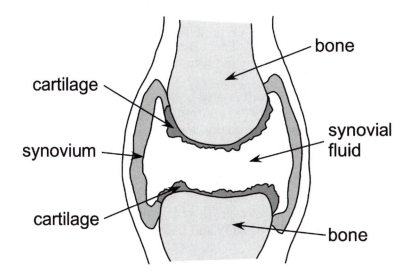

**Figure 13.2** *Simplified diagram of an inflamed joint*

associated with inflammation is quite clear and so NO is generally labelled as pro-inflammatory. If the production of NO could be suppressed it ought to effect an improvement in inflamed joints but the situation, particularly when it involves NO, is not that simple.

Joint tissue, such as cartilage, needs both food and oxygen and it is supplied with both, *via* the synovial fluid, by blood flow to the joint. So, factors influencing blood flow play a crucial role in maintaining the integrity of the joint and, of course, NO is one of these factors. It comes from the endothelial cells of vessels supplying the joint (eNOS) and from nerves (nNOS). The health and integrity of the joint is part of the natural defence against inflammation and so there is conflict in the roles played by NO in inflamed joints. NO is pro-inflammatory in that its production by iNOS is partly responsible for the inflammatory situation but NO from eNOS and nNOS activity is necessary for an adequate blood supply. Cartilage erosion, one of the most serious consequences of long-term joint inflammation, is probably due to overproduction of NO from iNOS. The three forms of NOS function in different parts of the joint and so the activity of iNOS may have no effect on blood vessel dilation and *vice versa*. Thus the administration of inhibitors of specific forms of NOS ought to have a profound effect on RA and initial experiments were very promising. However, some researchers have reported contradictory results. Part of the problem is that there are still no completely successful specific inhibitors of the different forms of NOS (see Chapter 17). A specific inhibitor of iNOS, which has absolutely no effect on eNOS and nNOS, ought to lessen the symptoms of inflammation but leave the blood supply to the joint unaffected. There is much activity in the pharmaceutical industry to this end but the results are, of course, commercially sensitive.

Thus inflammation is another area in which, at the moment, the role of NO appears to be somewhat contradictory, but this should be seen as a challenge and may, eventually, give us a very good drug for the treatment of RA. Current drugs do not address the problem of cartilage erosion, a feature of RA that could be arrested with a successful iNOS inhibitor. It might also be valuable in the treatment of osteoarthritis, where cartilage erosion is even more important. Before a clinically successful drug appears, however, there is a need for much more basic research on the chemistry of NOS and its inhibitors, and the biology of arthritis.

## FURTHER READING

L.N. Heiss, J.R. Lancaster, J.A. Corbell and W.E. Goldman, Epithelial auto-toxity of nitric oxide: role in the respiratory cytopathology of pertussis, *Proc. Natl. Acad. Sci. USA*, 1994, **91**, 267.

M. Stefanovic-Racic, J. Stadler and C.H. Evans, Nitric oxide and arthritis. *Arthritis Rheum.*, 1993, **36**, 1036.

J.S. Duffield, L.P. Erwig, X.Q. Wei, F.Y. Liew, A.J. Rees and J.S. Savill, Macrophages regulate resident cell apoptosis and mitosis in inflammed tissues. *J. Leukocyte Biol.*, 1999, 7.

K.J. Armour, R.J. van't Hof, K.E. Armour, D.M. Reid, L.M. Smith, F.Y. Liew, X. Wei and S.H. Ralston, Nitric oxide generated by the inducible nitric oxide synthase pathway plays an essential role in bone resorption. *J. Bone Mineral Res.*, 1999, **14**, 1115.

*Chapter 14*

# NO is Skin Deep

The skin is the largest organ in the human body. It protects us from heat, cold and the ultra-violet (UV) radiation in sunlight as well as being the first line of defence against germs of all kinds. Skin is continuously shedding dead cells from its surface, with new ones being generated from within. It consists of two major layers: the outer epidermis and the inner dermis (Figure 14.1). Beneath the dermis there is a layer of fat. The commonest cells in the epidermis are keratinocytes, which form at the base of the epidermis and, in the course of about two weeks, migrate upwards, become flattened and dehydrated, and are then shed from the surface. Melanocytes are located at the base of the epidermis and produce a protective pigment, melanin, which is distributed to surrounding keratinocytes. The dermis has a complex structure containing fibroblasts, structural protein (collagen), blood vessels, nerves, sweat glands, hair roots and muscles.

Because of its protective role, the skin should be considered part of the immune system, and so the presence of NO there is not surprising. When stimulated, for example by cytokines, the keratinocytes produce NO as well as hydrogen peroxide. Presumably this is to control of the numbers of pathogenic bacteria and fungi on the surface of the skin (of which there are very many), and kill any that enter the skin through a cut or abrasion. In a diseased condition such as psoriasis there is greatly elevated production of NO due to stimulation of iNOS. However, the exact role of NO in psoriasis is not fully understood, a situation made doubly difficult to study by the fact that the causes of psoriasis itself are not known. Is NO production one of the symptoms of psoriasis or an attempt by the immune system to control it? Effective treatments of the disease, such as corticosteroids and retinoids, inhibit iNOS activity, while other equally effective treatments, such as UV radiation

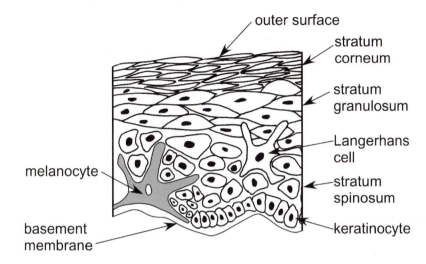

outer surface

stratum corneum

stratum granulosum

Langerhans cell

melanocyte

stratum spinosum

basement membrane

keratinocyte

**Figure 14.1** *Section through the epidermis of human skin*

and anthralin, increase iNOS activity. In cultured keratinocytes the response to NO is biphasic, in that at low concentrations cell proliferation is enhanced while at high concentrations there is greater cell differentiation. Clearly a healthy skin requires carefully controlled production of NO. Fibroblasts are very different from keratinocytes and respond differently to NO, suggesting that control of NO production must vary between different parts of the skin.

The external insult to which skin is most regularly subject is the UV radiation in sunlight. The advent of longer holidays and cheap mass travel have allowed dwellers in the temperate climes to expose themselves suddenly to massive doses of sunlight, often with disastrous consequences. Even the British summer sun can be dangerous. The short-term effect is vasodilation, with increased blood flow leading to reddening of the skin. The long-term effect is inflammation and the production of 'sunburn cells'. At a cellular level the effect of UV radiation on keratinocytes is the activation of iNOS for long periods, even after exposure to UV radiation has ceased. The initial vasodilation is the result of NO production. Sunburn cells are keratinocytes undergoing apoptosis or programmed cell death (see Chapter 12) and NO is one of the triggers for apoptosis. As keratinocytes proliferate and die within a short period of time, any disturbance of that process will manifest itself rapidly and severe sunburn is a serious matter. Why UV radiation causes enhanced NO production is not known for certain but

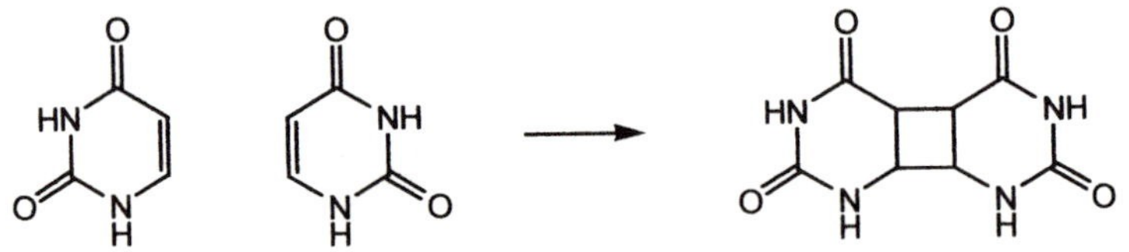

**Scheme 14.1**  *Cycloaddition reaction that occurs in DNA in response to sunlight*

UV radiation can bring about a whole range of chemical reactions and one of these must provide keratinocytes with just the stimulus they need to activate iNOS.

Even more worrying than sunburn is the now proven fact that UV radiation can cause skin cancers. UV radiation brings about chemical reactions in DNA and one cause of skin cancer is [2 + 2] cycloaddition to give thymine dimers (Scheme 14.1). This amounts to a mutation and, as a consequence, the genes that control cell growth and division go haywire. It does depend on skin type, but for many people the only certain safeguard against skin cancer is to protect the skin from sunlight with clothing. Traditional outdoor workers such farmers and game-keepers rarely strip when the sun shines; such foolishness is a characteristic of the urban dweller. Protective creams work only if they are frequently renewed. NO has a number of important functions once cancerous cells have started to grow. Firstly, cancer cells may display enhanced levels of iNOS activity and also have a modified or mutant form of the tumour-suppressor gene p53*. Cells with these two aberrant features grow rapidly, partly because of increased blood flow to the tumour, and inhibition of iNOS could have therapeutic value in such tumours. Secondly, NO encourages metastasis by lowering cell adhesion (see Chapter 2).

UV radiation is a major stimulus for human skin pigmentation and this explains why sunlight causes tanning. A tanned skin used to be seen as a sign of health but today, as skin cancers rise to epidemic proportions, current wisdom says otherwise. There is evidence to suggest that NO and cGMP production in keratinocytes is required for the generation of melanin from melanocytes. It could be that skin creams containing NO-releasing drugs are the tanning creams of the future.

As NO has such profound effects on cell growth and cell death, it is not surprising that NO is implicated in wound healing. The biphasic

---

\* A tumour-suppressor gene is one whose inactivation contributes to the development of a tumour.

effect of NO on keratinocytes mentioned above suggests that it may control key phases in the production of new keratinocytes, an essential part of skin regeneration. NO also regulates angiogenesis (the growth of new blood vessels) as the wound heals. Confirmation of the major role of NO in wound healing comes from the study of mice that have been genetically modified and do not have the gene for making iNOS. In these animals ('knockout mice'), which in other ways appear to be quite healthy, wound healing is significantly impaired. However, it still takes place, indicating that there are other substances in mouse skin that play roles parallel to that of NO. Since wound healing is such a complex process, involving accelerated cell growth and differentiation, the induction of iNOS and other mediators must be critically controlled for success. Even a minor disturbance, as occurs frequently with older people, may lead to a wound that does not heal.

The functions of NO in skin have proved to be many and varied. The early view that it was there to prevent infection has been overtaken by its role in controlling gene activity. NO can either enhance or shut down genes depending on complex factors surrounding the activity of transcription factors, substances that control the signalling ability of DNA. The story is far from complete, however, and there will probably be more surprises in the future.

We end this chapter on a simpler note, which shows that, although most of the chemistry of living processes is very complex, Nature does not spurn that which is simple. On the surface of human skin there is a source of NO that does not depend on the presence of any form of NOS. Sweat contains nitrite and its concentration may be as high as 3.4 µM. It is probably formed by the action of an enzyme (nitrate reductase), found in certain skin bacteria, on nitrate produced by the sweat glands as a way of excreting unwanted nitrogenous material. The skin is quite acidic and NO is produced from nitrite by protonation and hydrolysis (Scheme 14.2). This NO plays a part, along with NO from keratinocytes, in controlling the number of pathogenic bacteria on the skin. Such bacteria are a major cause of the spread of infections such as food poisoning and, in an age when washing your hands before eating is no longer *de rigueur*, NO from the skin is more necessary than ever. There is also a certain amount of nitrite (and therefore NO) in saliva,

$$2NO_2^- + 2H^+ \longrightarrow NO + NO_2 + H_2O$$

**Scheme 14.2** *Production of NO from skin nitrite*

because of the presence of nitrate reductase bacteria on the underside of the tongue (see Chapter 5), and so 'licking your wounds' and 'kissing it better' may not be such a bad idea.

## FURTHER READING

R. Weller, Nitric oxide – a newly discovered chemical transmitter in human skin. *Brit. J. Dermatol.*, 1997, **137**, 655.

D. Bruch-Gerharz, T. Ruzicka and V. Kolb-Bachofen, Nitric oxide and its implications in skin homeostasis and disease – a review. *Arch. Dermat. Res.*, 1998, **290**, 643.

*Chapter 15*

# NO in Nerves

The human brain is an organ of such complexity and delicacy that finding out how it works is one of the great challenges to science. The results of such study may impinge in a controversial way on philosophy and theology, and this makes it a particularly interesting area of research. With the aid of modern scientific techniques there has been much progress in elucidating some of the chemical reactions that occur in the brain but, before describing just some of these, we must look briefly at the cells that make up the brain.

The brain consists of a collection of nerve cells or neurones (Figure 15.1) that are highly specialized, supported by less specialized cells called glia. Neurones are nucleated cells with an elongated portion, the axon, along which electrical signals pass. These signals are not carried by electrons, which carry electrical signals in a metal wire, but instead by electrically charged ions. A normal human brain consists of about 1 000 000 000 000 neurones and neurones constitute about 2% of the total brain mass. New neurones are not normally created after birth and so all have to be produced during the nine-month gestation of the embryo, at the astonishing rate of 2.5 million per minute.

All neurones are independent and for the signal to pass from one neurone to the next it has to pass across a gap, or synapse, between neurones. The signal is carried across the synapse by a chemical called a neurotransmitter. When the signal reaches the axon terminal it stimulates release of a neurotransmitter, which then diffuses across the synapse. One very commonly occurring neurotransmitter is glutamate (**15.1**) which, on arriving at the membrane of the next neurone, binds to a receptor in the membrane. The best receptor for glutamate is the $N$-methyl-D-asparate receptor (NMDA; **15.2**). At NMDA receptors*

---

* So named because they also bind $N$-methyl-D-aspartate.

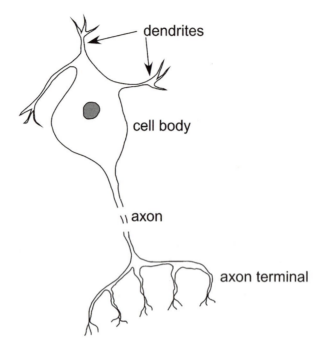

**Figure 15.1**  *Diagrammatic structure of a neurone or nerve cell*

                    15.1                                    15.2

glutamate opens calcium ion channels in the membrane of the neurone, calcium ions flow from the fluid in the synapse into the neurone and a new electrical impulse is set up that moves along the axon of that neurone. The movement of electrical signals along axons and across synapses is the physical manifestation of mental activity. Eventually the signal leaves the brain and continues its journey to a muscle, organ or gland along one of the peripheral nerves (see later), where it activates some response. However, the creation of a new signal in the postsynaptic neurone is not the only function of glutamate.

   There are many chemical processes occurring in the brain and one of them results in the formation of cGMP (see Chapter 1). In the early 1980s Takeo Deguchi, of the Tokyo Metropolitan Institute of

**Figure 15.2**  *John Garthwaite*

Neurosciences, noticed that cGMP formation in the brain requires arginine but was unable to explain the observation. In 1989 Salvador Moncada, who had recently demonstrated the arginine-to-NO pathway in the vasculature, suggested that Deguchi's observation indicated a role for NO in the brain and subsequently found direct evidence of NO-forming activity in brain tissue preparations. At about the same time John Garthwaite (Figure 15.2) and co-workers at Liverpool University showed that the effect of glutamate on brain tissue was the formation of a short-lived substance, which he later identified as NO. With this discovery the ascent of NO from a nasty pollutant to a life-sustaining chemical messenger molecule reached its zenith and animal physiology has not been the same since.

The presence of the enzyme NOS in brain tissue was soon demonstrated and the circumstances in which NO is generated are now well understood. Its formation occurs at the synapse and is linked to the behaviour of glutamate as a neurotransmitter, for NO is released in the postsynaptic neurone (Figure 15.3) when glutamate binds to an NMDA receptor. As mentioned already, binding of glutamate causes channels in the membrane to open and allow calcium ions from the fluid in the

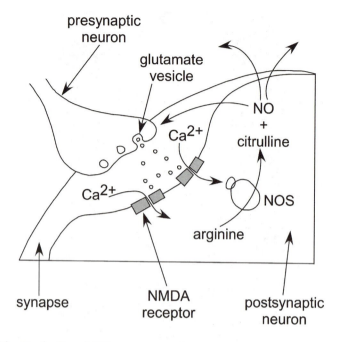

**Figure 15.3** *Production of NO as a retrograde messenger at a synapse*

synapse to enter the postsynaptic neurone and bind to calmodulin (see Chapter 3), which stimulates NOS in the neurone to convert arginine into NO and citrulline. The NO diffuses not only back across the synapse to the presynaptic neurone (for reasons that will be discussed shortly) but also to other neurones in the vicinity. Indeed Garthwaite has calculated that NO from a single source is likely to influence neurones over a large volume of brain tissue, perhaps large enough to contain a million synapses. NO can do this because it is produced in sufficient quantity, it is small enough to diffuse quite long distances rapidly and it is soluble in both the water and the lipid that make up most of the brain. Because of its unique chemical properties, NO in the brain behaves very differently from a neurotransmitter, whose sphere of influence is limited to one synapse.

The role played by NO in the brain is no simple matter and much has still to be elucidated; it is undoubtedly one of the many exciting areas of NO research. All three forms of NOS occur in the brain. eNOS is found in cerebral blood vessels and its role is thought to be much the same as that of eNOS in other blood vessels. iNOS is limited to glial cells and its function is still largely unknown. Most of the NO in the brain is from nNOS and is to be found in neurones, as described above. Why it is there is an amazing part of the NO story and the account requires a brief description of one aspect of brain function.

There is a phenomenon in brain activity called synaptic plasticity. If you keep stimulating synaptic activity in the same way the synapse becomes increasingly sensitive to that stimulus. The whole process is called long-term potentiation and is part of the learning process. It is one reason why reading a poem many times makes it easier to recall. Part of synaptic plasticity is thought to be due to a feedback mechanism involving a retrograde transmitter that passes from the postsynaptic neurone to the presynaptic one. Garthwaite suggested in 1989 that NO could be the retrograde transmitter and this is now generally accepted. In some way the fact that NO also reaches so many other synapses reinforces the signal further. Whether or not guanylate cyclase is the target for NO in long-term potentiation has yet to be established. As long-term potentiation appears to be part of the learning process we can claim that NO has a role in learning and memory. Equally, when memory fails, as in Alzheimer's disease, it may (and this is very speculative) be that production of NO is inadequate. It would be absurdly naïve to imagine that Alzheimer's disease could be reversed by NO-donor drugs but any future treatment will have to take into account the role of NO. The situation surrounding the onset of Alzheimer's disease is a complex one. There is evidence that its early appearance is partly genetic but there is also some evidence that a healthy diet of fresh fruit and vegetables, both of which contain high levels of antioxidants, can delay the onset. One reason could be that oxidant species, such as superoxide, remove NO by reaction but are themselves removed by the antioxidants in fresh food. There is also real evidence beginning to appear in the scientific literature to suggest that challenging mental activity, such as reading this book, keeps NO flowing healthily in the brain. Why not buy a second copy for the car?

Although NO has a crucial role in brain activity there are dangers in having NO in brain tissue. The most commonly occurring danger is that following a restriction in blood flow to the brain. This occurs after a stroke, when blood to the brain is interrupted either by atherosclerosis (furring up) of an artery or by a blood clot blocking an artery. Patients with thick or viscous blood, clotting disorders or inflamed arteries are particularly in danger of having strokes. The neural damage that results from a stroke is not primarily the result of oxygen starvation but due to the release of excessive amounts of the neurotransmitter glutamate. The glutamate concentration in the synapses may rise to $250\ \mu M$ if blood flow is restricted for any substantial length of time. This causes over-stimulation of the NDMA receptor and large amounts of calcium ions flow into the postsynaptic neurone through a membrane channel. Calcium ions activate several enzymatic processes, including the production of NO from arginine. There is direct experimental evidence for the production of high levels of NO in the brain of a rat when a key artery to the brain is occluded and, most significantly, this build-up can be prevented by giving the rat an NOS inhibitor before occlusion of the artery. The presence of excessive amounts of NO is harmful for reasons that have been described several times already. The fact that NO is part of the immune system, killing foreign bodies and cancerous cells, indicates clearly that it might, when present at a high level in an inappropriate situation, also kill the cells of the tissue that has produced it. There is always an optimum level of NO production for a healthy tissue.

The absence of oxygen in brain cells causes increased production of the enzymes associated with respiration, but they have no substrate. When oxygen finally reaches the brain on removal of the blockage, destructive oxygen radicals, such as superoxide, are formed. Also no NO can be produced by NOS from arginine in the absence of oxygen so there may be a surge of NO production when oxygen is readmitted to the brain, giving rise, on reaction with superoxide, to high levels of destructive peroxynitrite. This is why neural damage occurs not only during the stroke but also when reoxygenation occurs. It might be possible to lessen the damage by giving the stroke patient a dose of NOS inhibitor before removing the blockage. This would have to be done with care as the inhibitor would also cause constriction of the arteries leading to the brain, thus further restricting blood flow, a necessary condition for reoxygenation. The situation is further complicated by the fact that NO can inhibit the binding of glutamate to its receptor and so the production of further NO is prevented by NO itself. In that way NO can be said to have a protective role with regard to the brain.

In a book about NO it is easy to give the impression that NO does everything but it is important to stress that, in all these brain functions, NO is just one of many participants. We still have much to learn about the way in which NO interacts with other signalling pathways. Mice that have been genetically modified so that they cannot produce neuronal NO are ostensibly normal but abnormally aggressive. This rather startling observation shows just how little we understand the biochemistry of the brain, but of one thing we may be sure: NO is very important in brain activity. Exactly how important only time will tell.

Organs and muscles in the body are stimulated by messages from the central nervous system. The messages are carried to the organ or muscle by peripheral nerves and the mode of transmission along the peripheral nerves is very similar to that within the brain itself. Electrical signals pass along axons and are carried across synapses by means of neurotransmitters. Before describing the special role of NO in peripheral nerves we have to look a little more closely at the nature of activity within the nervous system. Our conscious thoughts and actions are only a tiny fraction of the total activity within the brain and peripheral nerves. If all consciousness were eliminated, vital physiological processes would continue virtually unchanged. This is due to the activities of the autonomic nervous system. This system regulates many bodily functions, including cardiovascular, respiratory, digestive, excretory and reproductive functions, without involvement of the conscious mind. It also regulates body temperature in a very efficient manner. There are two subdivisions of the autonomic nervous system:

1. *sympathetic*, which enhances tissue metabolism, enhances alertness and prepares the body to deal with emergencies ('flight or fight activities')
2. *parasympathetic*, which conserves energy and promotes sedentary activity like digestion ('rest and repose activity').

Both the sympathetic and parasympathetic systems target organs or muscle systems and the ultimate synapse links the peripheral nerve to the target organ or muscle. This synapse is known as the neuroeffector junction and the neurotransmitter here is crucial. It varies from one set of nerves to another. In cholinergic nerves the neurotransmitter is acetylcholine (**15.3**) and in adrenergic nerves it is norepinephrine (**15.4**). One of the characteristics of a neurotransmitter is that it has a very short *in vivo* life. Once it has moved across the synapse and activated the next neurone, or the organ if it is a neuroeffector junction, it is destroyed by enzyme action. This is necessary to prevent stimulation continuing when it is no longer relevant and to prevent the neurotransmitter diffusing into other synapses and stimulating the wrong nerves. The significance to the NO story of this characteristic of neurotransmitters will be clear shortly. The discovery of chemicals that block the neurotransmitter activity of acetylcholine and of epinephrine

$$CH_3CO-O \diagdown N^+(CH_3)_3$$

**15.3**

**15.4**

demonstrated that some vitally important nerves use neurotransmitters other than the two already mentioned, but the chemical nature of these agents defied elucidation. The nerves in which they were found were given the somewhat uninformative name of nonadrenergic noncholinergic or NANC nerves. Such nerves are found in the gastrointestinal, respiratory and urogenital tracts, as well as in certain blood vessels, and are a vitally important part of the autonomic nervous system.

A particularly straightforward NANC nerve for study is that supplying the retractor penis muscle of the bull (the BRP muscle) and much of the work of John Gillespie and William Martin of the University of Glasgow on the neurotransmitters of NANC nerves uses this particular muscle. It is anatomically fairly simple and readily available, along with NANC nerve supplying it, from slaughter houses. They attempted to extract the elusive NANC nerve neurotransmitter from BRP muscle by the use of dilute aqueous acid and a material (known as the inhibitory factor) was obtained that was an excellent mimic of the innervation of the muscle occasioned by the nerve. To register the inhibitory factor as a neurotransmitter caused problems as, from the ease of extraction from a postsynaptic muscle, it is clearly more stable *in vivo* than other neurotransmitters. Identifying the chemical nature of the inhibitory factor was unsuccessful until it was noted that the inhibitory factor not only brought about relaxation of BRP muscle but also the muscle of isolated blood vessels. At about this time (early 1980s) Robert Furchgott recognized the EDRF (see Chapter 1) and it appeared possible that the inhibitory factor and the EDRF were one and the same thing. With the identification of the EDRF as NO, the identity of the elusive neurotransmitter of NANC nerves became clear. It too is that ubiquitous, diatomic molecule. Thus two lines of research, from vascular and nerve physiology, came together, partly because William Martin went to work with Robert Furchgott in New York, and NO took on yet another biological role. NANC nerves were renamed nitrergic nerves in honour of NO. They contain nNOS for the generation of NO and the role of NO from nitrergic nerves is, as it is vascular smooth muscle, to activate guanylate cyclase and effect conversion of GTP into cGMP.

Soon there was more direct evidence that NO was the NANC nerve neurotransmitter. The NOS inhibitor $N^G$-nitro-L-arginine (**15.5**) was found to block the relaxation of the BRP muscle on innervation by the nerve supplying it. Also, visualization of the production of NO by electrical stimulation of a NANC nerve is now possible. NO is rarely the only neurotransmitter in a nitrergic nerve and it is probably more correct to refer to it as a co-transmitter.

15.5          15.6                    15.7

We are now in a position to explain why Gillespie and Martin were able to isolate what appeared to be the NANC nerve neurotransmitter (the inhibitory factor) from BRP muscle. NO released by the nerve at the neuroeffector junction diffuses into the muscle and some of it, on oxidation to $N_2O_3$, reacts with tissue thiols to give *S*-nitrosthiols (see Chapter 3):

$$RSH + N_2O_3 \rightarrow RSNO + HNO_2 \tag{1}$$

This then decomposes slowly to give NO:

$$2RSNO \rightarrow RS–SR + 2NO \tag{2}$$

The nitrosothiols most likely to be formed are *S*-nitroso-L-cysteine (**15.6**) and *S*-nitroso-L-glutathione (**15.7**).

Not all the experimental evidence is consistent with NO as the nitrergic nerve neurotransmitter as many chemical agents that destroy NO fail to block, or have only a minor effect on, nitrergic nerve neurotransmitter. For example, superoxide reacts rapidly and irreversibly with NO:

$$O_2^- + NO \rightarrow ONOO^- \rightarrow NO_3^- \tag{3}$$

Certain superoxide generating systems, however, distinguish clearly between authentic NO and the nitrergic neurotransmitter. Haemoglobin, which binds NO very strongly and is routinely used to see if a biological effect is due to NO, also has a slower effect on muscle relaxation evoked by nitrergic transmission than observed in other cases where NO is the messenger molecule. Because of these two, and other similar, observations, it has been proposed that the nitrergic neurotransmitter is not NO itself but an NO-releasing substance and the prime candidate is an *S*-nitrosothiol, such as **15.6** or **15.7**. Certainly these two substances will produce an effect on muscles that are normally nitrergically innervated. The most compelling evidence against *S*-nitrosothiols as the nitrergic neurotransmitter is the effect of copper ions. $Cu^+$ ions are highly effective as catalysts for the release of NO from an *S*-nitrosothiol (see Chapter 3) and enhance the relaxation brought about by *S*-nitroso-L-cysteine, but have no effect on the relaxation brought about by the natural nitrergic neurotransmitter.

The explanation for the lack of effect of superoxide-generating systems on nitrergic transmission appears to be the high levels of the enzyme superoxide dismutase (SOD) in muscle tissue. This effectively destroys superoxide before it can react with NO:

$$O_2^- + O_2^- + 2H^+ \rightarrow O_2 + H_2O_2 \tag{4}$$

If SOD is inhibited in some way superoxide-generating systems have a dramatic effect on nitrergic transmission. The distinctive effect of haemoglobin on nerve-evoked relaxation is due to the size of the haemoglobin molecule and it takes some time to diffuse into a thick piece of smooth muscle.

It is now widely accepted that the L-arginine-to-NO pathway accounts for nitrergic transmission at many sites and that the neurotransmitter released is NO. The transmission system differs from that in other nerves as the neurotransmitter is not stored in vesicles, as acetylcholine is, but synthetized on demand. As with so much about NO it has to be different. How NO as a neurotransmitter produced one of the most successful drugs of recent time is told in the next chapter.

## FURTHER READING

J. Garthwaite and C.L. Boulton, Nitric oxide signalling in the central nervous system. *Ann. Rev. Physiol.*, 1995, **57**, 683.

D.S. Bredt, P.M. Hwang and S.H. Snyder, Localization of nitric oxide synthase indicating a neural role for nitric oxide. *Nature*, 1990, **347**, 768.

E. Tarkowski, Å. Ringqvist, L. Rosengren, C. Jensen, S. Ekholm and Å. Wennmalm, Intrathecal release of nitric oxide and its relation to final brain damage in patients with stroke. *Cerebrovasc. Dis.*, 2000, **10**, 200.

D.W. Choi, Nitric oxide: foe or friend to the injured brain. *Proc. Natl. Acad. Sci. USA*, 1993, **90**, 9741.

J.S. Gillespie, J.C. Hunter and W. Martin, Some chemical and physical properties of the smooth muscle inhibitory factor in extracts of the bovine retractor penis muscle. *J. Physiol.*, 1981, **315**, 111.

A. Bowman, J.S. Gillespie and P. Soares-De-Silva, A comparison of the actions of endothelium-derived relaxant factor and the inhibitory factor from the bovine retractor penis on rabbit aortic smooth muscle. *Brit. J. Pharmacol.*, 1986, **87**, 175.

N.P. Wiklund, H.H. Iversen, A.M. Leone, S. Cellek, L. Brundin, L.E. Gustafsson and S. Moncada, Visualization of nitric oxide formation in cell cultures and living tissue. *Acta Physiol. Scand.*, 1999, **167**, 161.

*Chapter 16*

# The Truth About Viagra

Viagra is one of the most successful drug launches in recent years and its advent has occasioned more media comment than afforded to most new drugs. Its discovery was partly good science and partly good fortune, and it provides therapy for a condition which, although widespread, is difficult to treat. The male penile erection is the result of smooth muscle relaxation and blood vessel dilation in a structure known as the corpus cavernosum. Any number of visual, tactile, olfactory and imaginative stimuli prompt relaxation and dilation. These stimuli are processed in the brain and spinal cord and pass, *via* the peripheral nervous system, to the penis using a number of neurotransmitters, one of which is NO (see Chapter 15). Thus NO, acting as a neurotransmitter, arrives at the corpus cavernosum and the enzyme responsible is, of course, nNOS. Another probable source of NO is the vascular endothelium within the corpus cavernosum utilizing eNOS and there may also be role for iNOS. The net result of the presence of so much NO in the corpus cavernosum is the substantial dilation of all surrounding blood vessels and a massive flow of blood into the penis. Direct proof of the role of NO in the penile erection comes from the observation that topical application of glyceryl trinitrate may have the required effect.

When the connection between NO and the penile erection was announced in the scientific press certain sections of the media picked it up. It is probably the only time the phrase 'small, diatomic molecule' has appeared in *Cosmopolitan*. However, interest was shortlived as something far more newsworthy appeared, an oral drug for the treatment of erectile dysfunction. That the action of this drug was related to the role of NO caused little interest.

When guanylate cyclase is activated by NO the result the enzyme brings about is conversion of guanosine triphosphate (GTP) into cyclic guanosine monophosphate (cGMP). This conversion initiates a cascade of processes, leading to smooth muscle relaxation. The reversal of this process, resulting in vessel contraction and restriction of blood flow, occurs when cGMP is hydrolysed to guanosine monophosphate (GMP), a reaction catalysed by a set of enzymes called the phosphodiesterases. This set of reactions is common to all places in the body where NO is responsible for smooth muscle relaxation. In the 1990s the pharmaceutical company Pfizer started clinical trials of sildenafil citrate (**16.1**), a phosphodiesterase inhibitor, which the company had developed as a possible treatment for high blood pressure. If cGMP hydrolysis is prevented then blood pressure can be maintained at a lower level because of sustained dilation of the blood vessels (Figure 16.1). Clinical trials showed that sildenafil citrate was not very effective as a treatment for high blood pressure and higher doses were given. At this stage some

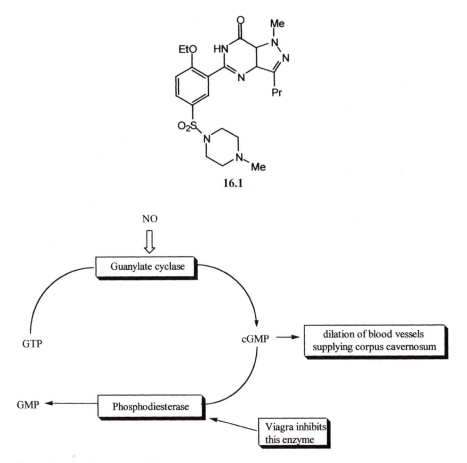

**16.1**

**Figure 16.1** *The action of Viagra*

of the men involved in the clinical trial reported a dramatic effect on penile erection. As penile erection involves the flow of blood into the corpus cavernosum it is easy to see how this should be the case but why sildenafil citrate should target that particular part of the vasculature is not understood. Once the reported observations had been confirmed Pfizer realized they had a brand new treatment for a condition that is quite widespread. The company renamed the drug Viagra and it has become a blockbuster (a term used in the pharmaceutical industry for a drug that sells well). Although it is used quite legitimately for the treatment of a condition common among men suffering from some forms of diabetes, as well as other illnesses, some might see Viagra as a lifestyle drug and its availability on an NHS prescription poses a moral dilemma for doctors. It is also available without prescription on the Web, which is worrying as taking it does have adverse side effects and these might be serious for some men.

The story of the development of Viagra as a treatment for erectile dysfunction is a good illustration of the role of serendipity in drug development. Serendipity is the faculty of making fortunate discoveries by accident and was coined by the novelist Horace Walpole in the eighteenth century from a Persian fairytale, *The Three Princes of Serendip*, in which the heroes possessed this gift. There are many examples of serendipitous discovery of drugs, most notably tranquillisers, which were discovered during a search for muscle relaxants. One of the reasons why the development of new drugs takes so long and is so expensive is that it is so difficult to predict that a certain chemical structure will have the required therapeutic effect. With modern computing techniques and vast databases listing the biological activity of millions of compounds the situation may improve over the coming years but there will always be a place for serendipity in drug discovery, a view that the Viagra team at Pfizer would endorse.

## FURTHER READING

J. Rajfer, W.J. Aronson, P.A. Bush, F.J. Dorey and L.J. Ignarro, Nitric oxide as a mediator of relaxation of the corpus cavernosum in response to non-adrenergic, noncholinergic neurotransmission. *New Eng. J. Med.*, 1992, **326**, 90.

K.E. Andersson, Pharmacology of penile erection. *Pharmacol. Rev.*, 2001, **53**, 417.

Pfizer, http://www.viagra.com

# NO From NOS – Detractors and Stimulants

Biological enzymes operate on a lock and key principle. The protein of an enzyme in its active state is folded such that the reactant, or substrate, fits into a specific volume of the enzyme (the active site) and is held there while the reaction takes place. The products are then released, allowing another molecule of the substrate to bind to the enzyme. An enzyme inhibitor is a substance that binds to the enzyme but does not then undergo the reaction for which the enzyme is a catalyst. Arginine (**17.1**) is the natural substrate for NOS and once the arginine has bound to the enzyme it reacts to form citrulline and NO. An NOS inhibitor binds to NOS, but citrulline and NO are not subsequently produced. Clearly the presence of an inhibitor lowers the amount of NO formed by the enzyme. Some important drugs used in medicine are simply enzyme inhibitors. One example is penicillin, which inhibits the enzyme catalysing the chemical reactions involved in the construction of bacterial cell walls. It is because of weakened cell walls that the bacteria die. Inhibitors are also important tools for investigating the mechanism of enzyme action. For research purposes the most commonly used inhibitor for NOS is *N*-monomethyl-L-arginine (**17.2**), always referred to by its initials L-NMMA. There are others, but most are other derivatives of arginine. L-NMMA is an inhibitor for all the forms of NOS, although there are claims that it is a slightly better inhibitor of eNOS than of the other two forms. One of the great challenges to the pharmaceutical industry is to produce inhibitors that are truly selective between the forms of NOS. An example would be an inhibitor that could completely inhibit iNOS, say, but not affect eNOS or nNOS at all. This has yet to be achieved, which is not surprising. In the three isoforms of the enzyme, eNOS, nNOS and iNOS, there is only

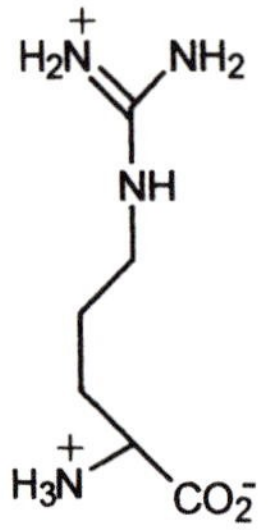

17.1

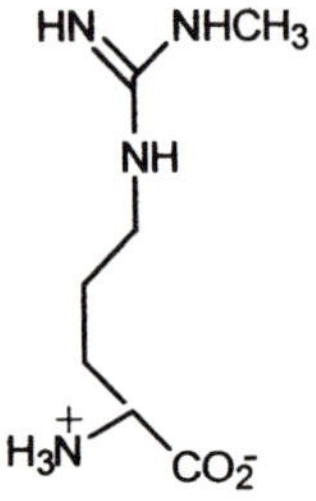

17.2

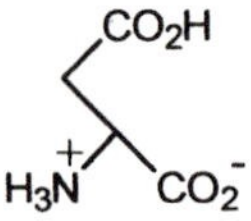

17.3

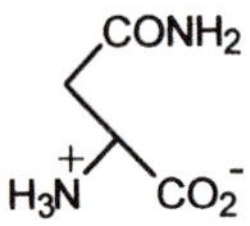

17.4

one major difference in the amino acid sequence of the active site of the oxygenase domain (see Chapter 3). Aspartic acid (**17.3**) is found at position 376 in the long protein chain of iNOS and nNOS, whereas it is asparagine (**17.4**) in eNOS. Finding a substance that is sensitive to such small differences is a challenge but selective inhibitors are very important as possible drugs for the treatment of rheumatoid arthritis (see Chapter 13).

Patrick Vallance and his colleagues at University College London have found that there is a naturally occurring inhibitor of all forms of NOS, asymmetric $N,N$-dimethyl-L-arginine (**17.5**), again known by its initials, L-ADMA. Its presence in a cell seriously diminishes the amount of NO formed by NOS. L-ADMA is formed by the enzymatic methylation of arginine in a protein and subsequent hydrolysis of the protein to its constituent amino acids. These processes are very active in growing animals, and there the levels of L-ADMA achieved could become high enough to compromise seriously the production of sufficient NO for cardiovascular health. But Nature, however, has developed a cure. There is a naturally occurring enzyme, dimethylarginine dimethylaminohydrolase (DDAH), which converts L-ADMA into citrulline and dimethylamine (Scheme 17.1).

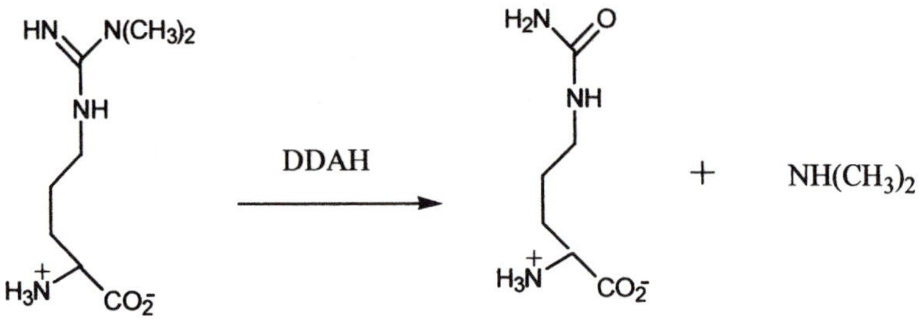

DDAH

$+$     $NH(CH_3)_2$

17.5

**Scheme 17.1**  *Conversion of L-ADMA into citrulline and dimethylamine catalysed by the enzyme DDAH*

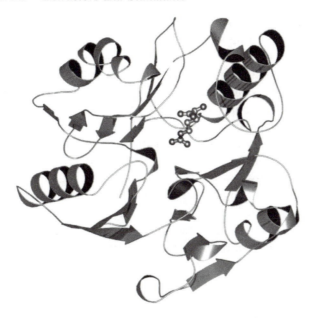

**Figure 17.1** *Structure of the enzyme DDAH with citrulline bound to the active site as determined by X-ray crystallography*

The structure of DDAH has been determined (Figure 17.1). It has four thiol groups in its active site and they are readily converted into disulphide by oxidation, *e.g.* by superoxide, leading to inactivation of the enzyme. Inactivation of DDAH causes raised levels of L-ADMA, a condition associated with a number of diseased states in humans, including high blood pressure, high cholesterol levels and heart failure. This is unsurprising given that L-ADMA is a NOS inhibitor, and reduced NO levels could lead to impairment of artery function and to coronary heart disease. So the role, superficially, of DDAH appears to be the removal of an adventitious and unfortunate inhibitor, which might be present at dangerous levels. However, it could have a role in controlling NO levels, particularly NO from iNOS. When the amount of NOS induced turns out to be too high, leading to dangerously high levels of NO in tissue, it may be that L-ADMA is transported to the site and acts as a rapid inhibitor of iNOS. L-ADMA binds *reversibly* to NOS; it is therefore said to be a reversible inhibitor of NOS. This means there is competition between L-arginine and L-ADMA for the active site of NOS, and L-ADMA can be displaced from the enzyme if the L-arginine concentration is raised. This is the theoretical basis for giving supplementary arginine to patients suffering from mild coronary heart disease. There is good clinical evidence that L-arginine, given daily in doses of around 5 g for some weeks, improves coronary blood flow.

If there are inhibitors of NOS, are there also substances which activate NOS, giving humans a healthy boost of NO? The answer appears
to be yes. The Kuna Ameridians, living on islands off the coast of
Panama, do not suffer from age-related increases in blood pressure,
although they eat a high salt diet, which is thought to be dangerous.
However, if Kuna people move to a city, they develop the same levels
of blood pressure as any urban population and so it follows that something in their native diet protects them from age-related blood pressure
increases. The good news for some is that the protection comes from
eating cocoa, the raw material of chocolate. The Kuna people consume,
on average, five cups of cocoa a day as well as adding cocoa powder to
many of their recipes. They continue doing this if they move to Panama
City but, instead of using home-grown cocoa beans, they use commercial cocoa and the good effect is lost. Commercial cocoa powder has
gone through a number of processes – fermentation, drying, roasting
and alkalizing – and these lower the procyanidin content of the cocoa.
Procyanadins are a family of flavanoids and one member is the
epicatechin dimer B2 (**17.6**). Another distinctive feature of the Kuna
people is that their urine contains higher than usual amounts of nitrite,
a distinction which is also lost when they cease eating native cocoa
beans. This observation led scientists to examine the effect of flavanoids
on NOS activity and some flavanoids were found to be strong activa-
tors of the enzyme. Eating native cocoa beans therefore gives you
flavanoids that activate NOS, producing a healthy shot of NO to keep
your blood pressure down and, at the same time, raising nitrite levels in
urine. Sadly, much of the goodness has gone out of cocoa by the time it
is turned into chocolate but no doubt native cocoa beans will soon be
available in health food shops.

**17.6**

17.7                                    17.8

Another naturally occurring material that may be an activator of NOS is garlic. That there are benefits of garlic to cardiovascular health is widely believed. It is said to decrease blood pressure, inhibit platelet aggregation, reduce blood fat and cholesterol, and to possess antibacterial and antifungal properties. There is good scientific evidence for these claims. The main odiferous component of *Allium sativium* is a sulphur-containing compound, allicin (**17.7**), which has beneficial properties, but there is evidence to suggest that a minor component, ajoene (**17.8**), is even more potent as an inhibitor of platelet aggregation. Naturally occurring ajoene is a mixture of the *E* and *Z* forms. Many of the physiological properties of garlic parallel those of NO and so a link between the two was looked for. *In vitro* experiments showed that garlic extract activates NOS in isolated platelets but why this should be so is not clear. In view of the biological activity of *S*-nitrosothiols (see Chapter 4) the presence of so much sulphur in garlic components may provide the clue. If the disulphide in allicin or ajoene is cleaved to give two sulphides, then the formation of *S*-nitrosothiols from locally produced NO is possible. Whereas NO has only a short life *in vivo*, an *S*-nitrosothiol is a long-lived source of NO. A further medical role for garlic has been suggested. Premature birth is a troublesome and costly event, and garlic extract is particularly effective at activating NOS in placental tissue. The NO thus produced could relax uterine muscle and delay labour until full-term. Quite how much garlic a pregnant woman would have to consume to guard against premature birth is not clear but it might be a socially unacceptable amount. Much of the beneficial effect of garlic is lost on cooking so eating the raw clove is required. Raw garlic goes well in a salad.

## FURTHER READING

J. Leiper and P. Vallance, Biological significance of endogenous methylarginines that inhibit nitric oxide synthase, *Cardiovascular Res*, 1999, **43**, 542.

J. Murray-Rust, J. Leiper, M. McAllister, J. Phelan, S. Tiley, J. Santa Maria, P. Vallance and N. McDonald, Structural insights into the hydrolysis of cellular nitric oxide synthase inhibitors by dimethylarginine dimethylaminohydrolase, *Nature Struct. Biol.*, 2001, **8**, 679.

N.K. Hollenberg, G. Martinez, M. McCullough, T. Meinking, D. Passan, M. Preston, A. Rivera, D. Taplin and M. VicariaClement, Aging, acculturation, salt intake, and hypertension in the Kuna of Panama. *Hypertension*, 1997, **29**, 171.

E. Block, S. Ahmad, J.L. Catalfamo, M.K. Jain and R. Apitz-Castro, Antithrombotic organosulfur compounds from garlic: structural, mechanistic and synthetic studies. *J. Am. Chem. Soc.,* 1986, **108**, 7045.

F.G. McMahon and R. Vargas, Can garlic lower blood pressure? *Pharmacotherapy*, 1993, **13**, 406.

I. Das, N.S. Khan and S.R. Sooranna, Potent activation of nitric oxide synthase by garlic: a basis for its therapeutic applications. *Curr. Med. Res. Opin.*, 1995, **13**, 257.

# Chapter 18

# Why Does Soil Evolve NO?

NO appeared in the atmosphere long before humans ever lit a fire or drove a car. Not only is it produced in lightning strikes, it is also released by a number of micro-organisms in soil. Both processes are important parts of the nitrogen cycle, which is summarized in Figure 18.1. One small part of the cycle is of particular interest here. The micro-organisms that release NO into the soil do so whilst converting nitrate from biomass into nitrogen gas (or, in rare cases, $N_2O$). This process is called denitrification. NO is an intermediate in the denitrification process. Most of the NO produced during denitrification is chemically reduced and only very small amounts are released.

Denitrification may be looked upon in two ways. Either it is the final step in balancing the nitrogen cycle or it is a nuisance without which nitrogen fixation would not be necessary. The process is summarized below:

$$NO_3^- \rightarrow NO_2^- \rightarrow NO \rightarrow N_2O \rightarrow N_2 \tag{1}$$

The majority of denitrification is carried out underground or in marine muds. Here, where oxygen is scarce, there are bacteria that can use nitrate instead of oxygen for their respiration. However, not all denitrifying bacteria are in anaerobic environments. In Chapter 14 mention was made of denitrifying bacteria on the skin that use the enzyme nitrate reductase to produce nitrite. In this chapter we look at the way in which NO is formed, and how it is subsequently converted into $N_2O$ in the course of denitrification.

NO is produced from nitrite in a process catalysed by the enzyme nitrite reductase (NIR):

$$NO_2^- + e^- + 2H^+ \rightarrow NO + H_2O \tag{2}$$

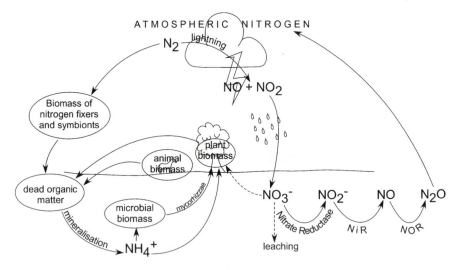

**Figure 18.1**   *A simplified schematic of the Nitrogen Cycle*

There are two genetically unrelated forms of nitrite reductase. One form contains iron in a haem structure, whilst the other contains copper. The copper in this second form is not contained within a porphyrin structure. All denitrifying bacteria use one or other of these forms, but the distribution between families of bacteria is not yet understood. The iron-containing form of NIR is a water-soluble enzyme and several members of the family from different denitrifying bacteria have been purified. It is a dimer, containing two different haem structures per subunit (Figure 18.2). The first of these is a so-called haem c structure.

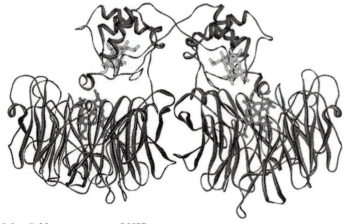

**Figure 18.2**   *Ribbon structure of NIR*

**Figure 18.3** *Structure of the haem d1 of NIR*
(Reprinted with permission from *Chem. Rev.*, 2002, **102**, 1201. Copyright 2002, American Chemical Society)

It accepts electrons into the enzyme from electron donors such as soluble cytochrome c. The second haem group is close to the first, with an Fe–Fe distance of about 20 Å and an edge-to-edge distance of about 10 Å. The porphyrin is rather special in that it contains fewer double bonds than normal and is designated d1. It is unique in biology to date, requiring a special biosynthetic process to assemble it. The porphyrin surrounding the iron centre contains fewer double bonds than usual as two bonds within the pyrrole rings are saturated. Also, there are two additional keto groups and one acrylate peripheral group (Figure 18.3). The polar peripheral groups provide extra places for the haem to attach itself to the rest of the protein structure *via* electrostatic interactions. More importantly, saturation of two pyrrole bonds in the porphyrin makes it more susceptible to out-of-plane deformations, which alter the physical properties of the haem. The different electronic properties of the d1 haem will affect other ligands present, so NO behaves differently here than when co-ordinated to a regular iron porphyrin. The d1 haem is the site of nitrite reduction, and the proposed mechanism is summarized in Scheme 18.1.

The nitrite ion co-ordinates to the iron *via* its nitrogen atom, following which an oxygen atom combines with protons from two nearby histidines (at positions 345 and 348 on the enzyme) to release water. The result is an iron nitrosyl complex, which has been shown to contain iron(III) bound to neutral NO. This bond is rather fragile since iron(III) does not back-bond (see Chapter 9) and the NO is easily released. A conveniently situated tyrosine group, Tyr-25, facilitates this release

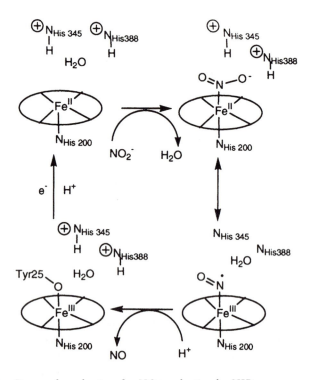

**Scheme 18.1** *Proposed mechanism for $NO_2^-$ reduction by NIR*
(Reprinted with permission from *Chem. Rev.*, 2002, **102**, 1201. Copyright
2002, American Chemical Society)

of NO. The oxygen of the tyrosine competes with the NO for the
co-ordination site on the iron, and wins because it forms a more stable
complex.

We can see a reaction very similar to this in a test-tube. In acidic
solution the nitrite ion oxidizes iron(II) to iron(III):

$$Fe^{2+} + HNO_2 + H^+ \rightarrow Fe^{3+} + NO + H_2O \qquad (3)$$

This is one laboratory synthesis of pure NO (see Chapter 6). Observers
of the reaction will note a transient brown coloration, due to the intermediate complex $[FeNO(H_2O)_5]^{2+}$. Nature has harnessed a well-known
reaction but the enzyme must somehow return to the starting state and
this is where the second haem group becomes involved.

Once the NO has been released and the tyrosine has co-ordinated to
the iron, an electron is transferred from the c haem to the d1 haem. The
tyrosine ligand is then replaced by a water molecule, and the catalytic
cycle is ready to begin again. It is incredible that the enzyme can

maintain such a delicate balance. Studies of NIR *in vitro* are dogged by build-up of a stable iron(II) nitrosyl, formed either by premature reduction of the iron(III) nitrosyl or recombination of NO with the starting iron(II) species.

The copper-containing NIRs are more biologically varied than those containing iron, even though they only account for about a third of the total denitrifying species. An X-ray structure of a copper NIR from *Achromobacter cycloclastes* shows that it contains three identical subunits. Each subunit contains two copper atoms, each of which is in a different chemical environment. The type 1 copper atoms are thought to be where the electrons enter the enzyme, whilst the type 2 copper atoms are the site of nitrite reduction. Enzymes that have been depleted of only their type 2 copper atoms show no catalytic activity. The type 1 and type 2 copper atoms are separated by about 13 Å but linked by adjacent amino acids in the same protein chain. This covalent linkage may help the transfer of electrons from the type 1 copper to the type 2 copper.

The actual mechanism of nitrite reduction in copper-containing enzymes is unclear as yet, partly because of conflicting evidence. Copper forms many stable nitrosyl complexes, most of which contain Cu(I) bound to NO. Complexes containing Cu(II) bound to NO are far less stable, so it is likely that NO is released on formation of such a complex in the catalytic cycle. It has been reported that the oxidized enzyme, which contains copper(II), binds nitrite bound *via* the oxygen atoms. A mechanism has been proposed that fits with this style of nitrite binding but it involves release of oxygen-bound NO, something yet to be observed elsewhere in metal nitrosyl chemistry. Also, this mechanism involves the participation of specific amino acids from the enzyme and when these amino acids are swapped for ones that cannot participate in the reaction, the enzyme is still 60% effective. Other workers have shown that copper(I) binds nitrite via the nitrogen atom, and concluded that nitrite is more likely to co-ordinate to a Cu(I) centre during the catalytic cycle.

Nitric oxide reductase (NOR) carries out the next stage in the denitrification process; it reduces NO to $N_2O$. The NORs are all part of a family of enzymes called the haem–copper oxidase superfamily. The best-known members of the group are the cytochrome c oxidases, which act as the final electron receptor in the aerobic respiratory chain, converting oxygen into water. They are mentioned in Chapter 19, in connection with the role of NO in the activity of fireflies. It is currently believed that the cytochrome c oxidases and the NORs both evolved

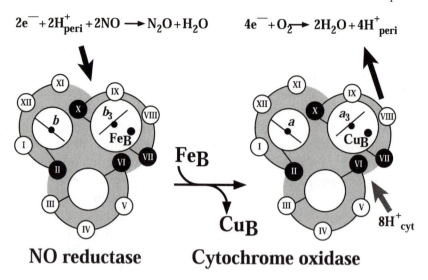

$$2e^- + 2H^+_{peri} + 2NO \longrightarrow N_2O + H_2O \qquad 4e^- + O_2 \longrightarrow 2H_2O + 4H^+_{peri}$$

**NO reductase**      **Cytochrome oxidase**

**Figure 18.4** *Comparison between NOR and cytochrome oxidase*
(Reprinted with permission from *Chem. Rev.*, 2002, **102**, 1201. Copyright 2002, American Chemical Society)

from an ancient NO-reducing enzyme. Both enzymes are membrane-bound, that is they are attached to membranes in the mitochondria of cells. Structurally, the two are very similar, but where the cytochrome c enzymes contain a copper atom, the NORs contain an iron atom. This iron atom is in a unique, non-haem environment (Figure 18.4).

Three types of bacterial NOR have been characterized to date. They are classified according to the substances used as electron donors. The most studied use cytochromes or small blue copper proteins as their electron source and consist of two subunits. They are present in all denitrifying bacteria and are designated cNOR. Another group consists of just one subunit and obtains electrons from quinols, hence the designation qNOR. A third type, $qCu_ANOR$, also obtains electrons from a quinol, but consists of at least two subunits and also contains copper.

All of the NORs face the challenge of getting two molecules of NO together and joining the nitrogen atoms to form a nitrogen–nitrogen bond. A mechanism for this has been proposed, based on studies of fully oxidized and reduced cNOR *in vitro*, and spectroscopic observations of the active site. It involves an oxo-bridge between two Fe(III) centres. As the iron centres are reduced, this bridge is weakened, water is evolved, and the haem iron atom remains five co-ordinate by co-ordinating to an adjacent neutral histidine. On co-ordination of the NO, however, the bond to the histidine breaks in a reaction analogous to that between NO and guanylate cyclase (see Chapter 1). Now two Fe(II)–NO bonds are formed. Such complexes should be stable, but here it is proposed that the two NO ligands are so close that they somehow react with one another, leading to evolution of $N_2O$ and regeneration of the bridged dimer (Scheme 18.2).

An alternative mechanism for the reaction has been proposed, in which only the non-haem iron co-ordinates to NO, and the $N_2O$ is formed *via* a dinitrosyl complex. The problem with this mechanism is that it does not provide a function for the adjacent haem iron centre, apart from electron transfer. Elsewhere in living things it seems probable that NO is reduced to $N_2O$ by only one iron centre. A class of denitrifying fungi has been discovered that use a copper-containing NOR. NOR in fungi does not bind to a membrane as it does in the bacteria and higher organisms. Instead, the NOR is water soluble. It consists of a single subunit and is a member of the so-called cytochrome P-450 family. Most cytochrome P-450 enzymes catalyse the mono-oxidation of hydrocarbons, but this one, called P-450nor, catalyses the reduction of NO to $H_2O$ and $N_2O$ using (unusually) NADPH as an electron donor. Although the mechanism of the reaction is still a matter for speculation, P-450nor only contains a single haem, so the reaction must be catalysed by one iron atom alone.

Discovery of denitrifying activity in fungi (and also in yeasts) was unexpected; originally it had been thought that denitrification was exclusive to prokaryotic (lacking a cell nucleus) organisms. Components of the denitrifying chain have also turned up in some interesting places. Let us turn once again to the overall denitrification process as shown in equation 2. Although denitrification is a means of utilizing nitrate to obtain energy as part of the respiratory cycle, the organism actually utilizes only the energy released in the first stage of the reaction, the conversion of nitrate to nitrite. In all the later stages, the energy released is either wasted or lost as heat. It would seem that these later stages are a means of ridding the organism of the toxic products of nitrate metabolism. Small (but measurable) amounts of NO are released into the surroundings during denitrification. This poses a real hazard to farmers making silage. If the grass used is too nitrogen-rich, denitrifying bacteria can thrive during fermentation, generating a lethal concentration of NO in the silo. As far as the organism is concerned, however, the amount of NO lost to the atmosphere is not nearly enough. Any organism unable to metabolize NO quickly after its production would poison itself. In organisms that use cNOR, all of which are denitrifyers, a gene regulator ensures that equal amounts of NIR and NOR are produced. This is not the case with organisms that use qNOR.

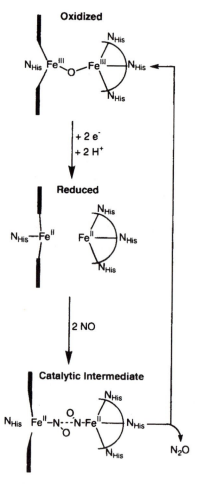

**Scheme 18.2**  *Proposed mechanism for reduction of NO by cNOR*
(Reprinted with permission from *Chem. Rev.*, 2002, **102**, 1201. Copyright
2002, American Chemical Society)

One such organism is *Ra. eutropha*. This is a denitrifyer, but it has
been demonstrated that here the production of NIR and NOR are not
linked. The surprise came with the discovery of qNOR in organisms
that do not denitrify and lack genes coding for NIR. Why would such
an organism need to produce qNOR? Well, many of these organisms
are pathogens that routinely invade mammalian cells. As we see in
Chapters 11 and 19, production of NO is one of the host's defences, so
a pathogen that can destroy NO quickly will have an advantage. Other
pathogens can also produce nitrate and nitrite reductase in addition to
qNOR. Perhaps in the case of these pathogens the ability to metabolize
nitrate when oxygen is not available may aid survival.

Until now this chapter has been concerned mostly with bacteria, but plants also play a major role in the nitrogen cycle. Experiments have shown that many plants may release NO in small, but measurable, quantities into the atmosphere. The amount of NO released is directly related to the amount of nitrate-containing fertilizer fed to them, and hence the concentration of nitrate in the plant cells. After considerable controversy it was shown that the NO released is actually produced by a soluble nitrate reductase, anomalously using nitrite rather than nitrate as a substrate. Reduction of nitrite is far less efficient than reduction of nitrate. This may be a way in which plants compensate for an excess of nitrate, although it may be frustrating to farmers and soil scientists who see their carefully applied fertilizer disappearing as NO!

In 2000, a plasma membrane-bound form of a nitrate reductase was found. This discovery was closely followed in 2001 by discovery of a membrane-bound nitrite reductase. This is especially exciting because the enzyme that produces reactive oxygen species in plants, NADPH-oxidase, is also found in plasma membranes. If the two act together then both NO and reactive oxygen species could be produced at the same time to give peroxynitrite, much the same as in the mammalian immune response. The role of NO in plant defence is also mentioned in Chapter 19.

Besides the role of NO in the immune system, there are other ways in which NO functions in plants. It has a profound effect on plant growth and development, amongst other things stimulating seed germination. It will be exciting to see the mechanisms by which NO carries out these functions and others yet to be discovered. NO research in animals has been thriving for more than 15 years but NO research in plants and microbes is only just beginning.

## FURTHER READING

I.M. Wasser, S. de Vries, P. Moënne-Loccoz, I. Schroder and K.D. Karlin, Nitric oxide in biological denitrification: Fe/Cu metalloenzyme and metal complex NOx redox chemistry. *Chem. Rev.*, 2002, **102**, 1201.

J. Hendriks, A. Oubrie, J. Castresana, A. Urbani, S. Gemeinhardt and M. Saraste, Nitric oxide reductases in bacteria. *Biochim. Biophys. Acta*, 2000, **1459**, 266.

W.J. Paynr, M.-Y. Liu, S.A. Bursakov and J. Le Gall, Microbial and plant metabolism of NO. *BioFactors*, 1997, **6**, 47.

P. Rockel and W.M. Kaiser, NO production in plants: nitrate reductase versus nitric oxide synthase. *Progr. in Botany*, 2002, **63**, 246.

*Chapter 19*

# NO is Everywhere

From what has been written about the biological role of NO so far it might be thought that the activity of NO is limited to mammalian physiology, but this is not the case. A few examples, from the hundreds that could be given, will illustrate the range of biological processes in which NO plays a part.

Locomotion is a fundamental requirement of almost all animals but the rate at which it develops after birth or hatching depends on the lifestyle of the animal concerned. Grazing animals, such as deer, which are prey to predators, can walk or even run within hours of birth but creatures that are hunters rather than hunted have a much slower development. A domestic cat is almost immobile for days after birth and the complex motor activity involved in hunting develops slowly over several months. Much has been learnt about the development of motor activity through a study of the development of swimming activity in tadpoles by Keith Sillar and his colleagues at the University of St Andrews. The anatomy of the neural system of the tadpole of the frog *Xenopus laevis* is much simpler than that of higher animals. Because of this, it has been possible to identify the clusters of nerve cells in the brain stem of very young tadpoles that are concerned with locomotion (Figure 19.1). They contain NOS and produce large quantities of NO. With very young tadpoles there are short bursts of swimming, which cease when a solid object is hit and the tadpoles hang motionless, presumably to avoid predators, attached to the object by means of cement glands. This type of swimming disappears within 24 h of hatching and, as the animal begins filter feeding on tiny particles in the water, swimming becomes more rhythmic and is more purposefully controlled. During motion the swimming muscles are activated and co-ordinated movement is brought about by the cessation of muscle activity, first on one side of the tadpole and then on the other. There are special cells joining the two sides of the

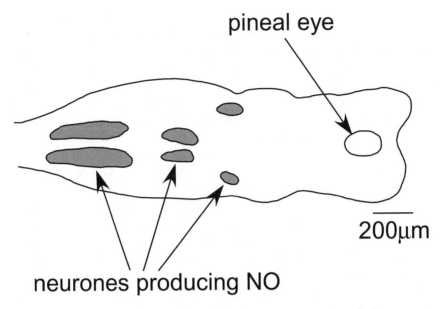

pineal eye

200μm

neurones producing NO

**Figure 19.1** *NO-producing neurones in the brain stem of the young tadpole*

spinal cord to ensure that inhibition passes from one side to the other. There are two inhibitors of muscle activity produced by tadpoles: glycine (**19.1**) and γ-aminobutyric acid (GABA; **19.2**). The messenger molecule that carries the message from the brain to the cells producing the inhibitors is, of course, NO, which originates in the clusters of nerve cells mentioned above. If tadpoles are placed in water that contains either a source of NO or an inhibitor of NOS activity, their swimming pattern is severely disrupted because NO increases the strength of the inhibition. Clearly tadpoles need an optimum amount of NO as much as humans do.

The horseshoe crab (*Limulus polyphemus*) is a very primitive animal that has remained unchanged for 500 million years. It has a very rudimentary vascular system but nevertheless cells of the vascular system contain NOS and produce NO. It might have been supposed that NOS would have evolved along with the sophisticated vascular system found in mammals but this appears not to be the case. NOS is one of the largest and most complicated enzymes known and so it is surprising to find it in such a primitive life form as the horseshoe crab. It is possible that in early evolution, when the world was a very different place, the role of

$H_3\overset{+}{N}$ $CO_2^-$

**19.1**

$H_3\overset{+}{N}$ $CO_2^-$

**19.2**

NO produced by NOS was very different from the current one and later higher animals took over NOS for control of the vasculature. The early atmosphere may have contained reactive oxygen species and NO production was necessary to protect an animal from these destructive species. The anatomical complexity of higher animals may be associated not so much with new and more complex enzymes, but with more elaborate and co-operative ways of using enzymes that already exist. Or it could be that we do not really understand the evolutionary process and an elaborate enzyme in a primitive life form is not odd at all.

The next examples come from the insect world. The bug *Rhodnius proxlixus* uses NO in a very disturbing way. In its salivary gland there is NOS, which produces NO to enhance vasodilation in its victim and prevent platelet aggregation as it feeds by sucking blood. From a rather unsavoury blood-sucking bug we turn to a more romantic insect. Fireflies, which are really beetles not flies, have long inspired poets* and delighted children. For scientists how and why they produce flashes of light has proved both fascinating and puzzling. The light comes from cells (photocytes) in an organ on the beetle's abdomen called a lantern where a chemical reaction (the luciferin–luciferase reaction, Scheme 19.1) that emits light takes place. Light emission is controlled by cells in the beetle's brain.

The purpose of firefly flashing is courtship. The flash is only a few milliseconds long but that is sufficient to attract fireflies of the opposite sex. Different species of firefly have different patterns of flashing and a male firefly uses the flashing pattern to ensure that he is courting a member of the same species. The immediate trigger for flashing is a sudden burst of oxygen supplied to the luciferin-containing structures within the photocytes by shutting off mitochrondrial respiration. Mitochrondria are subcellular structures (see Chapter 11) that consume large amounts of oxygen to provide the cell with energy. Nerve cells that activate the insect's lantern do not terminate at the photocytes, but on other cells some distance away. The activation of photocytes therefore requires a signal to pass from the endings of the nerve cells to the photocytes, a distance of about 17 μm. It is NO that fulfils this role. If you put fireflies in an observation chamber and give them a low dose of gaseous NO they become excited and flashing is more sustained and more rapid than is normal. What the NO does is to

---

\* Now sleeps the crimson petal, now the white;
  Nor waves the cypress in the palace walk;
  Nor winks the gold fin in the porphyry font:
  The fire-fly wakens: waken thou with me.
  *The Princess*, Tennyson

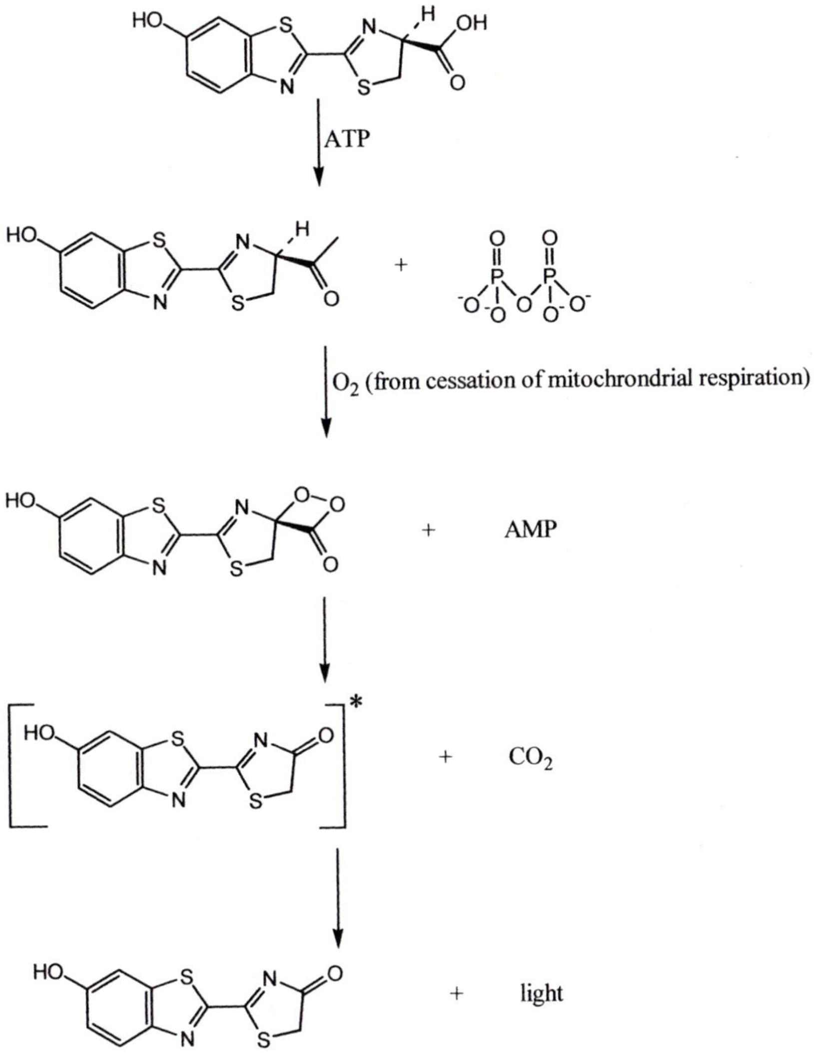

**Scheme 19.1** *Luciferase reaction responsible for the flashes of fireflies*

suppress mitochrondrial respiration within the photocyte transiently by inhibiting the enzyme cytochrome c oxidase. This process can be reversed by light and so the firefly's light may restore mitochrondrial respiration and thus terminate the flash. Insects do not have a vascular system and the effect of the NO produced by fireflies is limited to controlling its mating ritual.

NO occurs in plants not only as part of the nitrogen cycle (see Chapter 18) but also in the arginine-to-NO pathway. Plants are affected by pathogens as much as animals and infection in a plant triggers production of superoxide and hydrogen peroxide as toxic agents, but the toxicity of these two agents is too low to explain the cell death that occurs. There is now evidence that plants enhance toxicity by the same means as animals: the concomitant production from arginine of NO, which reacts with superoxide to give the highly toxic agent peroxynitrite. Addition of an NO-donor drug to a culture of plant cells infected by a pathogen results in greatly enhanced cell death. Equally, NOS inhibitors compromise, for example, the response of *Arabidopsis* leaves to the pathogen *Pseudomonas syringae* and actually promote disease and bacterial growth. At the same time, the NO produced by plants from arginine induces genes in the plant to produce protective natural products. Plants were producing their own pesticides long before humans thought of doing so. Whether NO has function within the plant kingdom other than that of protection from disease only time and much hard work will reveal.

Now that scientists are looking for NO it is popping up in some unexpected places. Soon a generation that has grown up with NO will be studying science and the element of surprise will have vanished but at the moment NO continues to astonish. We can conclude this chapter on what, for many people, will be a cheerful note. As has been observed in various places throughout this book, production of too much NO is harmful and may be the cause of some of our infirmities. St Paul offers some advice on this matter in his letter to Timothy: 'Drink no longer water but take a little wine for thy stomach's sake and thine often infirmities'.* Wine, and red wine in particular, is a good scavenger of NO and possibly the advice is more soundly based than was previously thought.

## FURTHER READING

D.L. McLean, S.D. Merrywest and K.T. Sillar, The development of neuromodulatory systems and the maturation of motor patterns in amphibian tadpoles. *Brain Res. Bull.*, 2000, **53**, 595.

M.W. Radomski, J.F. Martin and S Moncada, Synthesis of nitric oxide by the haemocytes of the American horseshoe crab (*Limulus polyphemus*), *Phil. Trans. Roy. Soc. London B*, 1991, **334**, 129.

J.M.C. Ribeiro, J.M.H. Hazzard, R.H. Nussenzveig, D. Champagne and F.A. Walker, NOS activity from a hematophagus insect salivary gland, *FEBS Letts*, 1993, **330**, 165.

* 1 Timothy 5, verse 23

B.A. Trimmer, J.R. Aprille, D.M. Dudzinski, C.J. Lagace, S.M. Lewis, T. Michel, S. Qazi and R.M. Zayas, Nitric oxide and the control of firefly flashing. *Science*, 2001, **292**, 2486.

M. Delledonne, Y. Xia, R.A. Dixon and C. Lamb, Nitric oxide functions as a signal in plant disease resistance. *Nature*, 1998, **394**, 585.

J.V. Verhagen, G.R.M.M. Haenen and A. Bast, Nitric oxide radical scavenging by wines. *J. Agric. Food Chem.*, 1996, **44**, 3733.

*Chapter 20*

# Reflections

The NO story, as it has so far unfolded, contains a number of contrasting examples of how the process of scientific discovery works. The popular idea that scientists work by the remorseless application of scientific logic and careful experimentation to a problem is, in most cases, wide of the mark. Scientific progress is generally much more haphazard than that.

Much of what we know about the natural world comes from simple curiosity. Early scientists, such as Priestley and Hales, mixed materials they had to hand simply to see what happened and this curiosity gave the world nitric oxide (Chapter 6). They had little or no idea of the structure of a molecule of nitric acid but noted that when the acid was poured over copper turnings a previously unknown gas was evolved. The structure of the gas was mysterious but they observed that it had certain characteristics, such as a low solubility in water, and produced another new, brown gas on contact with the atmosphere. These researchers had no recognizable target and would have been dismissed by the tabloid press of the day as 'useless boffins'.

Rather than curiosity it was serendipity that led to the discovery of the EDRF (Chapter 1). The origin of the word serendipity is described in Chapter 19 and it simply means the faculty of making fortunate discoveries by accident. In such discoveries chance plays a large part but it is only the great scientist who sees the significance of what chance has provided. In the words of one of the greatest scientists of all time, Louis Pasteur: 'Dans les champs de l'observation le hasard ne favorise que les esprits préparés'.* There is always scope for an inspired guess in scientific progress and that was, in large part, how the EDRF was identified as NO. With hindsight the identification is fairly obvious but hindsight

---

* Where observation is concerned, chance favours only the prepared mind.

is a wondrous thing; with hindsight we could all be geniuses. Anyone who makes a guess that is not correctly inspired, and publishes it, is in danger of ridicule and so that route to success and fame is a dangerous one.

Once the arginine-to-NO pathway in the vasculature had been established, recognizing it in other physiological processes, such as brain activity (Chapter 15) and the immune system (Chapter 11), was a matter of reading the scientific literature (no small thing these days) and doing some lateral thinking, an attribute not possessed by all scientists.

The development of the catalytic converter (Chapter 10), on the other hand, comes nearer to a conventional piece of science. The work was sharply targeted and success came from the use of the relevant chemical knowledge that had been acquired over the years by the work of other scientists. The converter is a beautiful piece of technology and there was little serendipity or guesswork in its development.

With the exception of the development of the catalytic converter, most of what has been described above is known as 'blue sky research': at the time it was not clear where it was going to lead. But now, with so much known about NO, the challenge is to turn that information into useful technologies, such as new drugs or better ways of removing NO from the environment. So far there has been surprisingly little progress here. Such developments require different skills from those used in blue sky research but the work is just as exciting. This part of the NO story has hardly begun.

# Subject Index